INSTANT
ENGINEERING

Portable Press
An imprint of Printers Row Publishing Group
10350 Barnes Canyon Road, Suite 100, San Diego, CA 92121
www.portablepress.com · mail@portablepress.com

Correspondence regarding the content of this book should be addressed to Portable Press, Editorial Department, at the above address. Author and illustration inquiries should be addressed to Welbeck Publishing Group, 20–22 Mortimer St, London W1T 3JW.

Portable Press
Publisher: Peter Norton
Associate Publisher: Ana Parker
Editor: April Graham Farr
Senior Product Manager: Kathryn C. Dalby
Produced by Welbeck Non-fiction Limited

Library of Congress Control Number: 2019946082

ISBN: 978-1-64517-054-9

Printed in Dubai

23 22 21 20 19   1 2 3 4 5

All illustrations provided by Noun Project with the exception of pg15 Romanvs Roman Mojzis/Shutterstock, pg37 Zvereva Yena/Shutterstock, pg41 Santio103/Ivan Feokistov/Shutterstock, pg84 Svetlana Maslova, pg93 Emre Terim, p96 Dn Br/Shutterstock, pg108 Naci Yavuz/Kawano/Shutterstock, pg109 Pulsmusic/Shutterstock, pg120 Dn Br/Shutterstock, pg131 Udaix/Shutterstock, pg135 Babich Alexander/Shutterstock, pg146 Molekuul_Be/Shutterstock, pg154 Dovla982/Shutterstock, pg155 & p163 Morphart Creation/Shutterstock, pg164 Caramelina/Shutterstock

INSTANT
ENGINEERING

KEY THINKERS, THEORIES, DISCOVERIES, AND INVENTIONS EXPLAINED ON A SINGLE PAGE

JOEL LEVY

**PORTABLE
PRESS**

San Diego, California

CONTENTS

TRANSPORT

BIOENGINEERING

AEROSPACE & ARMAMENTS

ELECTRICAL & COMPUTERS

MECHANICAL

INTRODUCTION

Today, engineering is defined as the application of scientific principles to practical purposes, and specifically to designing and developing systems, processes, structures, and machines to use the resources of nature to serve the needs of humankind.

The origins of engineering, however, lie not in science but in the down-to-earth, hands-on experience of resourceful individuals, making and building, learning what worked, and developing a body of practical knowledge.

The word *engineer* derives from the Latin root *ingenerare*, "to create": the same root as "ingenious." The first engineers were thus ingenious people, and the products of their ingenuity were engines; specifically, engines of war, for the origins of engineering as a profession and field lie in the military. Military engineers were called upon to design and build fortifications and the means to overcome them; weapons of war, especially large and complex ones such as catapults; and the infrastructure relied upon by armies and navies, from roads and bridges to get armies moving, to docks where ships could be constructed, to mines to bring down city walls. Many of these pursuits had obvious civil applications, and the development of civilization went hand in hand with the development of civil engineering: the building of canals and bridges, the digging of mines and quarries, and the construction of palaces and roads.

In ancient Greece, Rome, India, and China, a parallel tradition emerged of ingenious devices, artifices, and contraptions: machines. This tradition, known as mechanics, had important philosophical or academic aspects, but it was characterized by its practical nature, so that mechanics were defined, by themselves and others, in opposition to theoretical pursuits such as philosophy and science. Although today mathematics and science are central and crucial to engineering, historically many engineers distrusted this theoretical side. For those trained by apprenticeship, not in a school or university, and in America, for instance, this remained the case until as late as the mid-nineteenth century.

Elsewhere, however, a more rigorous conceptual and professional approach developed, and it was military engineering that once again was leading the way. The advent of gunpowder weapons increasingly demanded that the military engineer be able to apply scientific principles and mathematical precision, and in the early eighteenth century the powerful and advanced French state had created a school of military engineering to oversee artillery

and artillery-resistant fortifications, and a school of civil engineering to build the roads, bridges, and canals needed for an effective military. British and American engineering developed more in the commercial and private sphere, focusing on industrial processes and machines, but as these became more and more advanced and specialized, it became more and more important to apply scientific principles. Mechanical and agricultural engineering emerged as branches of engineering coequal to military and civil, and the march of technology and science would go on to add more branches to the engineering tree. Electrical engineering emerged in the nineteenth century, primarily driven by the telegraph industry. Chemical engineering arrived in the same century, while the twentieth century saw the development of nuclear, biological, space, and computer engineering. New fields continue to emerge, including genetic, geo-, and nano-engineering. Today engineering is arguably the preeminent field in both science and technology; the one with the greatest impact on humanity, from day-to-day life to the global economy, from the most mundane activities to the fate of the planet.

The way this book is arranged loosely mirrors the organization of engineering as a whole, with sections exploring the most important branches of engineering, including civil, biological, transport, military and aerospace, electrical, and mechanical. In addition, there are overarching or underlying concepts that do not fit within these fields, and some of these are treated in the first section. Stephen Hawking was once warned by his publisher that each equation he included in his book would halve his readership. With this dictum in mind, *Instant Engineering* mostly avoids introducing equations, or indeed any mathematics. But readers should be mindful that mathematics is the basic language of engineering, and that without the complex calculations of the engineer none of the immense edifice of modern engineering would function: bridges would collapse, mobile phones would not ring, laptops would overheat, and planes would not fly. Equally important are the scientific principles that underlie so much of engineering, and from these the book does not shy. In these pages you will meet the laws of thermodynamics, motion, conservation of energy, relativity, and many more besides, but all are explained and/or evoked in simple, accessible terms that help to provide

the reader with a deeper understanding of essential concepts and principles of engineering. Concepts and jargon are always defined or explained when introduced, and, in addition, a glossary at the end of the book provides brief definitions of key concepts and engineering terms.

Within each section, entries are arranged in more or less chronological order, and a timeline at the end of the book attempts to provide an overview of the chronology of the contents, although such an arrangement begs many questions. Even scientific and technological discoveries and advances can often not be pinned down to a particular time or person, and while engineering frequently overlaps with science and technology, it is even less of a clear-cut and individual pursuit. Definitively pinpointing the origins of "inventions" such as roads or domes is clearly impossible and indeed, as with many aspects of engineering, such concepts reveal the deep roots of engineering as a practice that stretches back to the origins of humanity itself. Defining features of our hominid lineage include tool use and tool making, the ability to shape and reshape our environment and to adapt nature to our priorities, rather than the other way around. These are key elements of engineering, and so in some ways we can consider engineering to be a core human competence: a major part of what it means to be human.

BEAM

*A beam is a basic structural element, longer than it is wide or thick, designed
to support loads and is itself supported at or near the ends.*

FORCES FROM ABOVE

The forces from the loads above are
generally vertical, due to **gravity**.
These forces generate forces of
compression, shear, and tension
within the beam.

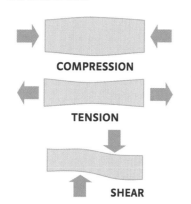

COMPRESSION

TENSION

SHEAR

MATERIAL CONSTRUCTION

In the past, beams were often
**made of oak, hewn from
a single trunk** into a square
section. Such beams
are still used in buildings today.
Modern beams are made of **steel
or reinforced concrete**.

INTERNAL FORCES

Internal forces produce **strains
and bending** of the beam.

TYPES OF BEAM

For both vertical
and horizontal
forces, **square-
section beams**
are best.

For forces in
all directions,
tubular beams
are strongest.

For **vertical forces**, the strongest
shape is an **I-beam**. For **horizontal
forces**, an **H-beam**, like an I-beam
on its side, is better.

Beam balances have been used
for weighing things for thousands
of years.

In a **beam engine** the
overhead beam is hinged in
the middle, and transfers the
force from the piston to the
pumping rods or other load
at the opposite end.

BIOENGINEERING

At the intersection between biology and engineering is bioengineering, also known as biomedical engineering, which ranges from sports science to agriculture, clothing to pharmaceuticals.

TIMELINE

- **c.11,000 BC** Prehistoric humans start to engineer biological fermentation systems to brew beer
- **c.950 BC** Prosthetic toe buried with ancient Egyptian mummy
- **c.700 BC** Etruscans craft false teeth from human and animal teeth
- **c.AD 1000** Medieval Europeans extract blue dye from woad on an industrial scale

- **1885** Prototype heart-lung machine developed in Germany
- **1941** George de Mestral returns from a walk with his dog and notices how plant burrs cling to its fur, giving him the inspiration for Velcro
- **1943** First dialysis machine
- **1954** British scientist and broadcaster Heinz Wolff coins the term "bioengineering" at the UK National Institute for Medical Research
- **1958** First implantable cardiac pacemaker
- **1959** Neurologist William Oldendorf gets the idea for a computer tomography X-ray scanner from watching a fruit-sorting machine
- **1963** Engineering in Medicine Laboratory created at Imperial College, London
- **1966** Biological engineering program created at University of California, San Diego
- **1972** First instance of genetic engineering with direct transfer of DNA from one organism to another
- **1985** Microfluidic engineering used to create popular pregnancy tester Clearblue
- **1997** Tissue grown around a biological scaffold produces Vacanti mouse
- **2012** Launch of Tricorder XPRIZE to stimulate development of portable diagnostic technology
- **2017** Thought-controlled bionic arm developed at Johns Hopkins
- **2019** Researchers at University of Illinois at Chicago create artificial leaf

IN THE MIX

Bioengineering is one of the ultimate **interdisciplinary** fields. As well as the fundamentals of biology and engineering, a student of bioengineering might expect to study **electrical and mechanical engineering, computer science, materials science, chemistry, nuclear physics, biochemistry, genetics, microbiology, agriculture, sports science,** and **medicine**.

BIOMIMICRY

An important principle in bioengineering is that **evolution,** working for billions of years through countless prototypes, **has achieved superior engineering solutions to many problems**. Copying or drawing inspiration from natural engineering is known as **biomimetic engineering** or **biomimicry**. A classic example is the invention of the fastener **Velcro** in the 1940s.

COLUMN

A column or pillar is a vertical member that transmits through compression the weight of the structure above to the surface below. They are the main vertical supports in millions of engineering structures: the compression members that support much of the built environment.

TYPES OF COLUMN

STONE COLUMNS

Stone columns may consist of a single piece of stone.

Stone columns usually have **cylindrical shafts**, with a **capital** on top and a base or **pedestal** below.

Taller columns were made of sections, often carved with central holes, and pegged together with stone or metal pins. **Nelson's Column** in London is made of **granite**.

The **Colosseum** in Rome, built between **AD 70 and 80**, consists mainly of stone columns.

STEEL, CONCRETE, AND BRICK COLUMNS

Modern columns are made of **steel, concrete, or brick**.

Small **metal or wooden supports are called posts**. Those with rectangular cross sections are called **piers**.

COLUMNS UNDER STRESS

Columns frequently fail during **earthquakes**, either by **bending or by twisting**. This failure may bring down the entire building that they were supporting.

Earthquake-resistant columns often have **vertical steel reinforcing bars** and horizontal ties within the rectangular concrete structure.

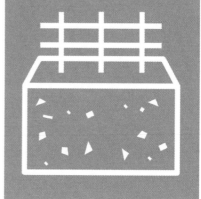

CONTROL THEORY

*Most dynamical engineering systems need control. For example a bicycle needs
a braking system to prevent disasters when going downhill.*

MODERN CONTROL SYSTEMS

ON   OFF

OPEN LOOP

The simplest form of **automatic
control** is an open-loop system.
A central heating system may
be programmed to operate from
8 a.m. to 10 a.m., whatever the
temperature of the building.

CLOSED LOOP

In a closed-loop system there is **feedback**
from the process. The central heating
system may be controlled by a **thermostat**,
which **switches the system on** when the
building is **colder** that the set temperature
of, say, **70°F**, and switches it off again when
the temperature reaches 70°F.

WATER CLOCK

One of the earliest known control
systems was the **clepsydra** or water
clock designed by **Ctesibius** around
250 BC. Water **dripped** from one
container into another, where the
depth of water indicated the time on
a **scale or dial**. To ensure a steady
flow, Ctesibius made sure the **upper
vessel** was always full to the **brim**.
These were the most **accurate**
clocks for hundreds of years.

CENTRIFUGAL GOVERNOR

Having greatly improved the steam engine around **1775**, **James Watt**
invented the centrifugal governor to control its speed. If the engine ran
too fast the **balls would fly apart**, and so slow the engine down to
a **sensible speed**.

In **1868** the mathematician
James Clerk Maxwell showed
mathematically how and why the
centrifugal governor was liable
to become unstable, responding
too slowly to events. Since then
control theory has become largely
mathematical.

AIRCRAFT CONTROL

In **1903** the **Wright Brothers** (see
page 108) only just managed to
control the **lift and stability** of their
aircraft, and all subsequent aircraft
have had **complex** control systems.

ELASTICITY

Elasticity is the capacity of a solid material to spring back and recover its shape after being deformed by a force. Engineers need to be familiar with the elasticity of the materials they use, in order to be able to calculate materials' behavior under stress.

HOOKE'S LAW

"As the extension; so the force" was how Robert Hooke described his 1675 observation, which came to be called Hooke's Law. The **extension of a spring is directly proportional to the extending force**.

Hooke's Law applies to most elastic materials for small deformations. The strain, or **relative deformation**, is proportional to the stress, or load. **Double the weight**, and the **spring will extend twice as much.**

ELASTICITY OF DIFFERENT MATERIALS

In **metals**, stress causes the **atomic lattice** to change shape slightly. When the stress is removed, the **atoms spring back** to their original positions.

In **rubbers**, the stress causes the **long polymer molecules** to stretch. Rubbers and similar **elastomers** are able to **stretch or bend much more than metals**.

DEFORMATION

Most **solid materials** will deform steadily until they reach a **yield point**.

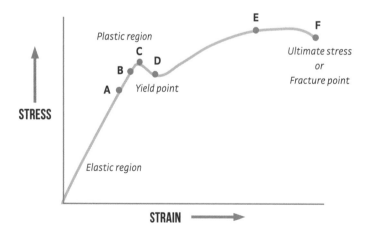

WORK-HARDENING

Many metals and plastics can be strengthened by work-hardening; some plastic deformation **changes the atomic structure** and **extends the elastic limit** of the material.

If more force is applied, the deformation becomes plastic and **irreversible**; the material will not revert to its original shape.

EMERGENCE

Emergence is the principle that systems can exhibit properties that result from the interaction of elements but that are not present or could not be predicted from those elements in isolation. In other words, the whole may behave in ways not expected from the sum of its parts.

SYSTEMS ENGINEERING

Emergence is an important principle in the field of engineering known as "**systems engineering**," which is the **study and practice of putting elements together so that they interact as a system**. An **engine** is a type of system; so is a **research department** or a **factory production line** or an **electricity grid** or a **pier** projecting into the sea.

GENERAL PRINCIPLES

ELEMENTS OF A SYSTEM

SYSTEM

TYPES OF EMERGENCE

Types of emergence include **simple, weak, and strong**:

Simple emergence, also known as **synergy**, is when the **elements** of a system **combine** to produce an effect that they cannot produce unless **working together**, but which **could be predicted or expected**.

$$A + B = C \checkmark$$

Weak emergence is a property that **might be expected** to arise from a system, but the **extent of which cannot be predicted** just by looking at the **sum of the system's parts**.

$$A + B = C \times ?$$

Strong emergence is when properties arise that **could not be predicted** and are **only seen when the system starts to operate**.

$$A + B = ?$$

EXAMPLES OF EMERGENCE

An **airplane's ability to fly** might be considered an example of **simple emergence**. On their own, the wings and other elements of an airplane cannot fly, but flight emerges as a property of the whole system. An example of **strong emergence** might be the problem that beset London's pedestrian **Millennium Bridge** on opening in June 2000, when it transpired that **minor oscillations**, with which the bridge was designed to cope, caused a **feedback loop** to generate **severe swinging**. Many engineers working in **artificial intelligence** pin their hopes on intelligence proving to be an emergent property of a sufficiently **complex neural network**.

ENERGY

Energy is the capacity to do work. Many great engineering projects need skillful manipulation of energy transfer, whether in the power train of a vehicle or the heat expansion of a bridge.

ENERGY AT WORK

Water at the top of a **waterwheel** has potential energy. As it falls under gravity it **pushes the waterwheel around** and so generates work.

Wind has **kinetic energy**, because the air is moving. It can turn the **blades of a windmill** or a wind turbine, and so convert the energy into work.

When one form of energy is converted to another, some **heat is always generated**. In **1798 Benjamin Thompson** noticed that when **cannons were being bored**, enough heat was generated to **boil water**.

JAMES PRESCOTT JOULE

In **1843** British scientist **James Prescott Joule** carried out a series of experiments to show that a particular amount of **mechanical energy** could be converted to a **precise amount of heat**.

He arranged for **falling weights** to **drive paddles** in a beaker of water, so that the **friction** that slowed the movement of the paddles **warmed the water**. The unit of energy, the **joule**, is named after him.

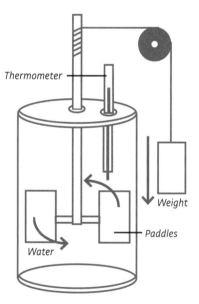

One joule is the **energy expended**, or the work done, in raising a weight of one **newton** (e.g., a small apple) through a height of **1 meter**.

RUDOLF CLAUSIUS

In **1850 Rudolf Clausius** realized that in any change no energy is gained or lost; the total energy **remains the same**, even if some is converted to heat. This is the first law of **thermodynamics**.

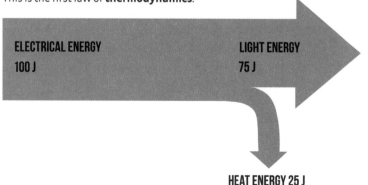

ELECTRICAL ENERGY
100 J

LIGHT ENERGY
75 J

HEAT ENERGY 25 J

FINITE ELEMENTS ANALYSIS

Finite elements analysis (FEA) is the technique of breaking a complex shape or structure into many small units (finite elements), which can be more easily mathematically described, and then linking together the calculations for these elements to achieve a mathematical description (or analysis) of the whole.

WHY USE FEA?

Mathematics is a **crucial tool** for engineers, who need to work out the physical properties of structures or objects and how these will change in response to, for instance, **loads or stresses**. But **complex shapes and structures**, and **complex processes** such as **fluid dynamics**, pose extreme mathematical challenges: the mathematics involved is impossible to work out without using **shortcuts or simplifications**. One such shortcut is the **finite elements method**. It was developed to enable **computer analysis** of the physical properties and responses of shapes and structures, and is particularly used in **simulation software**.

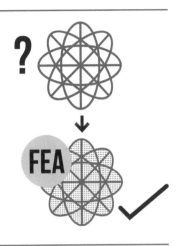

BASIC STEPS OF FEA

The first step, known as **discretization**, is to **divide up** the shape or structure into elements, or **discrete units**. These connect to one another at points called **nodes**. The complete assemblage of elements and nodes is known as a **mesh**. For each element, the **equations** describing its properties are specified, and then these equations are connected at the nodes, resulting in a **set of simultaneous algebraic equations**, the solution of which describes the whole mesh (at least approximately). Using this mesh, a **computer program** can now simulate the **properties and responses of the structure**—for instance, giving values for the **compressive forces** acting at any point in a **load-bearing column**.

TYPES OF FEA

FEA modeling can be used in three different types of analysis:

Static: For **modeling** with a **variable that remains constant**; e.g., modeling the bearing pad of a bridge.

Dynamic: For **analyzing** the **dynamic response** of a structure that experiences **dynamic loads**; e.g., modeling the impact of a human skull on a surface.

Modal: For **simulating responses to vibration**; e.g., an engine starting up.

GEOENGINEERING

Large- and even planetary-scale interventions to reduce greenhouse gases and limit or reverse global warming are known as geoengineering. There are two main categories of geoengineering: solar and carbon.

SOLAR GEOENGINEERING

Also known as **solar radiation management**, this category covers strategies to **reduce the amount of energy** from **solar radiation** that gets into the atmosphere, land, and oceans. The aim is to **counteract the warming effect** caused by **greenhouse gases** and related phenomena (such as diminishing ice caps). Possible techniques include:

Stratospheric aerosols: introducing tiny, **reflective particles** into the upper atmosphere to **reflect sunlight** back into space, mimicking the effect of a "**volcanic winter**."

Space parasols: putting **sunshades** or **mirrors** in orbit to **intercept sunlight** before it reaches the atmosphere.

Albedo enhancement: making either the **Earth's surface** or **clouds** more **reflective** to increase the proportion of **solar radiation** that is **reflected back into space**.

GLOBAL FRANKENSTEINS

There are many **objections to geoengineering**, but the biggest is the problem of **unintended consequences**. Critics argue that the worst way to counteract anthropogenic disruption of the biosphere is with **more disruption**.

CARBON GEOENGINEERING

Also known as **greenhouse gas removal**, this category includes techniques to take greenhouse gases such as **carbon dioxide** out of the atmosphere, or stop them getting there in the first place. They include:

- **Ambient air capture**: Taking greenhouse gas directly from the atmosphere using huge **"scrubbing" machines**, and then storing the gas in some form.

- **Ocean fertilization**: Adding **nutrients** to the ocean to encourage **blooms of plankton** and other **aquatic microorganisms**, which **incorporate carbon dioxide into their cells**.

- **Afforestation**: Global-scale **tree planting** to create **carbon sinks** in plant form.

- **Enhanced weathering**: Exposing vast amounts of **minerals** to soak up carbon dioxide by **reacting** with it to create **carbonate rocks**.

- **Adding alkali to the ocean**: Putting huge quantities of **alkaline minerals** such as **limestone** into the ocean so that it will **react with carbon dioxide** to create **carbonates** and **counteract ocean acidification**.

- **Carbon capture and sequestration**: Captures **greenhouse gases** emitted from **energy production and industrial processes** and locks them away before they can get into the atmosphere, but only reduces global carbon if the energy is produced by burning specially grown **biomass**.

- **Biochar**: **Charring biomass** to create **charcoal** and then burying it so that it cannot rot and release carbon.

INFORMATION THEORY

Information theory is a branch of mathematics that seeks to encode and transmit a piece of information in the most effective way.

SHANNON'S BITS

The idea was first proposed by **Claude Shannon** in **1948**, in a paper called "A mathematical theory of communication."

Shannon's vision unified all types of communication—**telephone signals, text, radio waves, and pictures**—which could be encoded in **binary digits or bits**.

00010010 00010010 00010010
10100110101 101001101 101001101
00010010 00010010 00010010
11100100111100100111100100 1
00010010 00010010 00010010

He showed how information could be **quantified** with **absolute precision**, and demonstrated the **essential unity of all information media**.

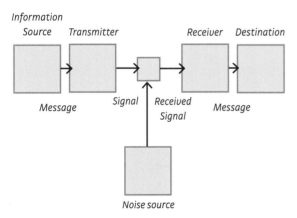

Information
Source   *Transmitter*   *Receiver*   *Destination*

Message   *Signal*   *Received Signal*   *Message*

Noise source

Shannon explained that all **communication systems** have a **common form**. **Noise** is likely to be **added** during **transmission**.

INFORMATION ENTROPY

This is a measure of the **uncertainty in a message**.

Sent over a noisy channel, a message may become a set of possible messages. The goal is for the receiver to **reconstruct the correct message**, with no errors, in spite of the channel noise.

Information theory leads to **precise error-free data** compression (e.g., ZIP files), and less precise data compression (e.g., MP3s and JPEGs).

GROUNDBREAKING INVENTIONS

Information theory was **vital to the success** of the **Voyager** missions to deep space, the invention of the **compact disc (CD)**, the introduction of **mobile phones**, and the **development of the internet**.

SPEED LIMIT

Every communication channel has a speed limit, the **Shannon Limit**, measured in **bits per second**. Try to exceed this limit, and **errors creep in**. Below this limit the message can be free of errors.

HEAT

Heat is a form of energy (see page 18)—the least useful form, because converting heat to useful energy or work is difficult and inefficient.

DEFINITION HEAT IS IN FACT THE RAPID MOTION OF ATOMS OR MOLECULES.

GENERAL PRINCIPLES

EARLY THEORY

In the **eighteenth century** heat was thought to be a **fluid called caloric** that flowed from a **hot object to a cold one**. Put a **hot rock** in a **bucket of water**, and "caloric" flowed from the rock to the water.

In **1798** the American-British-German spy-scientist-diplomat **Benjamin Thompson** (see page 18) noticed that **boring iron cannons** generated heat, and that blunt borers generated more heat than sharp ones. He reasoned that there was **no caloric in the borers or the iron** before boring began; so where did the heat come from?

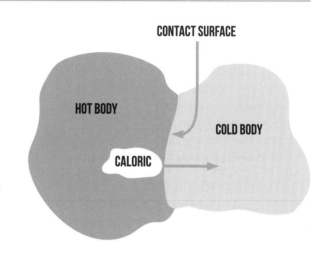

CONTACT SURFACE

HOT BODY

COLD BODY

CALORIC

FRICTION

James Prescott Joule reckoned that the heat came from **friction**, and noted many cases where this seemed to be the case. Eventually, after much ridicule and rejection, he got the **great physicist Lord Kelvin** to agree with him.

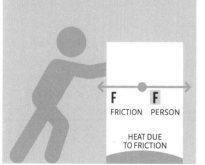

F F
FRICTION PERSON

HEAT DUE
TO FRICTION

WASTED ENERGY?

Heat is often just wasted energy, but it can be **converted to more useful forms**, for example in the **Stirling engine** (see page 168) or in **power stations** (see page 138).

LATENT AND SENSIBLE HEAT

In his **1847 lecture** "On Matter, Living Force, and Heat," Joule identified the terms **latent heat** and **sensible heat** as **components** of heat, each affecting distinct **physical phenomena**.

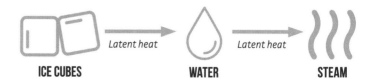

ICE CUBES   *Latent heat*   WATER   *Latent heat*   STEAM

LEONARDO DA VINCI

Leonardo da Vinci was an Italian painter, scientist, and engineer, born of unmarried parents in the town of Vinci, near Florence in Italy.

 DATES APRIL 15, 1452–MAY 2, 1519

ART

Painted *La Gioconda* (the **Mona Lisa**) and many other famous works.

Made beautiful **anatomical drawings of people and animals**, and wrote **extensive notes** in mirror writing.

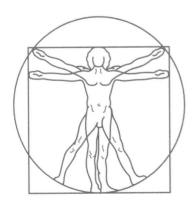

He made many **sketches of birds** and their **wings**, and how they **fly**.

INVENTIONS

Had an extraordinary gift for drawing technical illustrations, both of **existing machines** . . .

. . . and of **imaginary ones**, including several **flying machines**; no machine was able to fly until 250 years after his death.

An engineer is usually someone who builds things. Leonardo built a **series of barricades** to protect Venice from attack and, in 1502, he produced a drawing of a **720-ft. bridge for Constantinople**, but most of his sketches and drawings were **only ideas**. He had an exceptional vision of engineering and of how things worked, or **might work in the future**.

He designed a **bicycle** and a **parachute**, centuries before their time, as well as **musical instruments**, a **mechanical knight**, **hydraulic pumps**, **reversible crank mechanisms**, **finned mortar shells**, a **steam cannon**, and a **giant crossbow**.

LEVERS

A lever is a type of simple machine: a device for doing work by amplifying force, so that force exerted over a larger distance can be concentrated to act over a shorter distance. The ancient Greek mathematician and engineer Archimedes stated the law of the lever.

LAW OF THE LEVER

The law of the lever states in **algebraic terms** that the force multiplied by the **distance from the fulcrum** on one side of the fulcrum equals the same quantity on the other side:

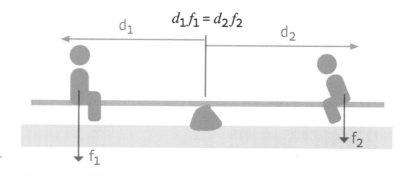

$$d_1 f_1 = d_2 f_2$$

- **Pre-20 million years ago** Primates first start using sticks as force amplifiers

- **c.5000 BC** Balance beam used by ancient Egyptians for weighing

- **c.200 BC** Archimedes proves the law of the lever

- **c.AD 1400** Renaissance scientists define the six basic types of machine: lever, wheel, inclined plane, screw, wedge, and pulley

LEVERS IN UNEXPECTED PLACES

Levers are found everywhere in the world and in many different guises. You are making use of the **law of the lever** when you use any of the following devices: **scissors, bottle openers, nut crackers, pliers, hedge shears, bolt cutters, crowbars, wrenches, claw hammers, see-saws, scales, rackets, bats,** and **wheelbarrows**.

PARTS OF THE LEVER

A lever has four elements:

- **Beam**: a structural element that can **pivot** or move on the **fulcrum**
- **Effort**: the **force** that is exerted by a person or machine on a lever
- **Fulcrum**: the point at which a lever **pivots** or **hinges**
- **Load**: the **object** that is **acted on** by the lever

LEVER CLASSES

There are **three classes of lever**, depending on the relative positions of the **effort, fulcrum,** and **load**:

Class 1: effort and **load** are on **opposite sides** of the **fulcrum** (e.g., **scissors, pliers**).

Class 2: effort and **load** are on the **same side** of the **fulcrum** with the **effort further away** (e.g., **wheelbarrow, nutcracker**).

Class 3: effort and **load** are on the **same side** of the **fulcrum** but the **effort** is **between the load and the fulcrum** (e.g., **tweezers, baseball bat**).

MECHANICS

Engineering mechanics, also known as applied mechanics, is a specific branch of the field of physics known as mechanics. It is the application of mechanics to real-world examples, such as how materials, components, and structures respond to applied forces.

ENGINEERING VS CLASSICAL MECHANICS

Mechanics is the first and fundamental branch of physics. In **classical physics** (i.e., physics that does not fall within the remit of relativity or quantum mechanics) it is the **science of bodies in motion or equilibrium**. Classical mechanics relies on **Newton's laws of motion**. Applied mechanics is where classical mechanics meets the real world. **Applied mechanics** further **breaks down into two** categories: **statics** and **dynamics**.

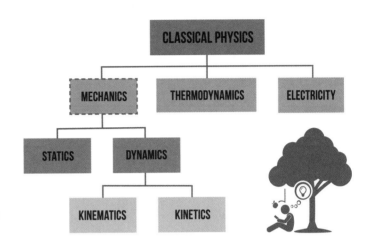

STATICS

This is the branch of mechanics concerned with **analysis of bodies in static equilibrium** (where all forces balance each other out). **Newton's second law of motion** says that force = mass × acceleration. Statics deals with situations where **acceleration** is **zero**, which means that the net force on a body must also be zero.

DYNAMICS

This is the branch of mechanics concerned with **analysis of bodies that are not static** (i.e., where net forces are not zero). It further subdivides into **kinematics** (analysis of a dynamic body *without* regard to the forces required for the motion) and **kinetics** (analysis of a dynamic body *with* regard to the motive forces).

FREE BODY DIAGRAMS

An important tool for engineers in statics is the free body diagram: a picture that shows **all the forces on a body at rest** (i.e., not under acceleration). All these forces must **cancel each other out**.

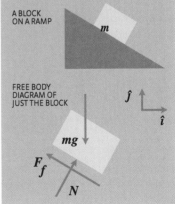

A BLOCK ON A RAMP

FREE BODY DIAGRAM OF JUST THE BLOCK

FLUID MECHANICS

There are other types of mechanics. **Quantum mechanics** is currently outside the remit of engineering, but areas like **fluid mechanics**, the science of fluids and bodies in **relative motion**, are important.

NANOTECHNOLOGY

Nano means very small. A nanometer is one billionth of a meter: 1nm = 10⁻⁹m.
Nanotechnology is engineering on the atomic or molecular scale.

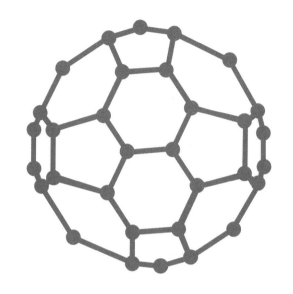

GENERAL PRINCIPLES

1980S RESEARCH

In **1981** the invention of the **scanning tunneling microscope** by **Gerd Binnig** and **Heinrich Rohrer** enabled scientists to see **actual atoms** and bonds between them, and later to manipulate individual atoms.

In **1985 Harry Kroto**, **Richard Smalley**, and **Robert Curl** discovered the extraordinary molecule C_{60}, in which 60 carbon atoms are arranged in a shape like a **soccer ball**. It was called **buckminsterfullerene**, after the structures of the architect Buckminster Fuller, later shortened to "**buckyballs**," or "**fullerene**."

NANOPARTICLES IN COMMON PRODUCTS

Nanoparticles of silver metal are **bactericidal**, and have been incorporated into the surface of **cutting boards**. Intel makes **microprocessors** by nanolithography.

Nanoparticles of titanium dioxide are used in **sunscreens**, and also added to **glass** so that it **repels dirt** and becomes **self-cleaning**.

PRESENT AND FUTURE RESEARCH

Nanotechnology is an **exciting and rapidly expanding** field of research. Scientists are experimenting with **molecular self-assembly**, which should enable complex chemical structures to be synthesized automatically, while **Martin Burke** has developed a molecular **3-D printer**.

FULLERENE

Fullerene is **too big a molecule** to be truly nano, but the later derivatives **carbon nanotubes (CNTs)** are around **1nm** in diameter and from 100nm to 0.5m long. They are extraordinarily strong, and are used in **boat hulls**, **car parts**, **sporting goods**, **stain-resistant textiles**, and many other places.

GRAPHENE

Graphene (**a single sheet of carbon atoms**) is used to make **plastics stronger** and **more conductive**, and in **electronic and photonic circuits**, **transistors**, and **solar cells**.

RENEWABLE ENERGY

Renewable energy is energy derived from resources that are not finite, but that are constantly renewed. Developing and scaling the technology to make renewable energy sufficient, cost-effective, reliable, and available is one of the great engineering challenges of our age.

c.3500 BC Sailboats use wind to drive ships

400 BC First written reference to the waterwheel

3rd century BC Greeks and Romans use concentrating mirrors to light torches with solar power

1st century AD Heron of Alexandria describes a wind-powered machine

644 Earliest known reference to a windmill, in Persia

1839 Edmond Becquerel discovers the photovoltaic effect

1879 First hydroelectric power plant built at Niagara Falls

1888 First wind turbine

1904 First geothermal electricity plant at Larderello, Italy

1954 First silicon photovoltaic cell

1966 First tidal power station on the Rance Estuary in Brittany, France

2000 First commercial wave energy system installed at Islay in Scotland

TYPES OF RENEWABLE ENERGY

Most but not all sources of renewable energy can ultimately be traced back to the **Sun**. The main categories are:

- **Solar**: using **solar radiation** to generate **electricity** directly, as with solar panels, or to generate **heat** that can be used directly or to generate electricity.
- **Wind**: wind turbines at sea or on land use blades to **convert wind energy into kinetic energy** to power electricity generators.
- **Hydro**: **Solar radiation** drives the **water cycle**, and **hydro power** taps into the power flows in this cycle, for instance by using falling water to turn **turbines**.
- **Ocean**: **Wave power** derives from **wind**; **tidal power** derives from the **gravitational effect** of the **Sun** and **Moon**; **temperature gradients** in the ocean derive from **solar warming** and currents; **currents** derive from a combination of other sources.
- **Geothermal**: Deriving ultimately from **decay of radioactive elements**, geothermal energy can be tapped directly for **heat**, or to generate **steam** for turbines.
- **Biomass**: Using **natural processes** to convert solar radiation to **chemical energy** provides a renewable source in the form of biomass that can be **burned** or turned into **biofuel**.

THE STORAGE PROBLEM

One of the greatest engineering challenges facing renewable energy is the problem of storing it so that it can be available where and when it is needed, and not just where and when the wind is blowing, where and when the Sun is shining, and so forth.

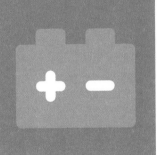

RISK

All engineering projects have an element of risk. The task of the engineer in any project is to assess all the possible risks involved, and try to allow for them in the design.

EXAMPLES OF RISK

Simple risk: When you wield a **hammer** you may hit your **thumb** instead of the **nail**.

Serious risk: Engineers may not allow enough in their **calculations** for building design for the structure to **withstand a powerful earthquake**.

FUKUSHIMA

On **March 11, 2011,** an unprecedented **tsunami** overwhelmed the **30-ft. seawall** at the Japanese town of Fukushima, and the **nuclear reactor** behind it, causing a major leak of **radioactive material**.

No one was killed immediately by the radioactivity but some **20,000 died from drowning** or other events following the tsunami. The sea wall was very high, but **should the engineer have built it higher?** The huge wall of **water was much more dangerous than the radioactivity**.

BRIDGES

On **December 28, 1879,** the Tay Bridge collapsed as a train crossed during a gale, causing **75 deaths**. The engineer, **Sir Thomas Bouch**, had been wrongly advised about **wind pressure**, and the cast iron used in the bridge was found to be **substandard**.

Bridges and similar structures are often deliberately **over-engineered**, allowing for much **greater loads** than are expected.

AUTOMOTIVE TRAVEL

Because they are in control of a car but are helpless in a plane, most people have poor judgment of the risk of each, and do not realize that **driving a car is far more dangerous than flying**. In the United States alone some **35,000 people die in car crashes** each year, while in the whole world, on average, **fewer than 1,000 die in plane crashes**.

SOIL MECHANICS

Buildings and civil engineering structures have to rest on something, and that something is usually soil. Soil is also an important component of structures such as embankments and dams. Understanding how the soil behaves in response to forces is crucial.

THE BIG SIX

Soil mechanics are determined by **six properties of soil**:

- **Friction**: How resistant to sliding is a mass of the soil? The more water, the less friction, so clay has lower friction than sands and gravels.
- **Cohesion**: **Attraction** between the particles that make up the soil stickiness. **Clay** is more **cohesive** than **soil** or **gravel**.
- **Compressibility**: How much will the soil compress under a **load**?
- **Elasticity**: To what degree will the soil **expand** back to its **previous density** after compression?
- **Permeability**: How well does **water flow** through the soil?
- **Capillarity**: To what extent is **water drawn up** from the **water table** through the soil?

FOUNDATIONS

Foundations are needed to **transmit the weight of a structure to the ground** in such a way that it remains stable. The **foundation type** must be **matched to the soil mechanics** or the soil may **shear** (slide along planes of movement) or **settle unevenly** under compression. Types of foundation include:

- **Spread-footing or pads**, where pads are placed directly beneath **load-bearing elements** such as **columns** or **walls**.
- **Mats**, where a **slab**, usually of **reinforced concrete**, **underlies the entire footprint** of a building.
- **Floating**, where **rigid box foundations** are set at such a depth that the **weight of soil removed equals the weight of the structure above**, so that the soil below ends up **bearing the same weight** as before.
- **End-bearing piles**, where **columns** extend all the way down to the **bedrock** to **transmit load directly**.
- **Friction piles**, which **transmit load to soil along their whole length**.

SLOPE STABILITY

Gravity exerts a **force** on the soil in a **slope**, which is **counteracted** by the forces of **friction** and **cohesion** between **soil particles**. **Soil engineers** grade the **stability** of a slope by the **ratio between these forces**. A ratio of precisely one indicates the forces are **perfectly balanced**. A ratio of two shows the **stabilizing forces are twice as strong as the motive force**. A slope that rates less than one is **likely to slip or collapse. Water levels and movement in the soil** will affect this ratio.

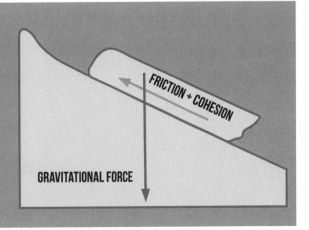

FRICTION + COHESION

GRAVITATIONAL FORCE

SURVEYING

Surveying means studying and measuring an area of land or a building, in particular identifying the corners of a piece of land or the plan of a building.

ANCIENT EGYPT

The **Great Pyramid of Giza** in Egypt was laid out with the sides lined up precisely **north-south** and **east-west**. The surveyors probably found due south by noting the **positions of sunrise and sunset** one day, and **bisecting the angle** between them.

ANCIENT GREECE

The Greeks were adept at surveying; the word *geometry* means **"measuring the land."**

Around **550 BC**, **Eupalinos** (see page 39) used simple surveying techniques to work out where to **start his second tunnel** on the island of **Samos**.

ROMAN TIMES

The Romans used simple techniques to plan **straight roads**, usually taking the **shortest route from A to B**.

The primary tool of the Roman surveyor, or **gromaticus**, was the **groma**, invented in ancient **Mesopotamia**. This was a simple **portable** instrument with **plumb bobs and right angles**.

SURVEYING TOOLS

In **1571** English scientist **Leonard Digges** described the **theodolite**, an instrument to measure horizontal angles, which is still used today.

In addition to the **theodolite**, **modern surveyors** use **measuring tape**; **total station**, which measures angles and distances; **3-D scanners**; and a **level**, an **optical instrument** used with a **leveling staff or rod**.

MAPPING

For mapping, both **aircraft and satellites** have been used, but in the UK the Ordnance Survey hopes to use **high-altitude solar-powered drones** in the near future, which can cruise for months in the **stratosphere** without needing a **battery change**.

THERMODYNAMICS

Thermodynamics is the branch of science that deals with heat, temperature, energy, work, and the relationships among them.

LATENT HEAT

In **1761 Joseph Black** discovered latent heat: the heat that is necessary to **melt ice**, and the **heat given out when steam condenses** to water.

WATT'S EFFICIENCY

In **1769 James Watt** had the brilliant idea of using a **separate condenser** to improve the efficiency of the **Newcomen steam engine** (see page 49).

CARNOT CYCLE

In **1824 French engineer Nicola Leonard Sadi Carnot** was keen to improve the efficiency still further.

He described the operation of an imaginary perfect **leak-free** steam engine—a heat engine operating between **two heat reservoirs**, with steam converting heat energy into **mechanical work**. This "Carnot cycle" earned him the title **Father of Thermodynamics**.

FIRST LAW

In **1850 Rudolph Clausius** showed that heat won't flow from a **colder object to a hotter one**. He also realized that energy is conserved in any operation of a **closed system**; it cannot be **created or destroyed**. This is the **first law** of thermodynamics.

SECOND LAW

In any real system, some energy is always **dissipated** as heat; this loss is **entropy**. This is the **second law** of thermodynamics. The fact that entropy is **always increasing** is one reason why **time runs forward**.

LORD KELVIN

In **1854 William Thomson** (later Lord Kelvin) and **William Rankine** formulated precise versions of the first and second laws.

Thomson invented the idea of an **absolute temperature scale**. This is measured in **kelvins**; zero K is the absolute zero of temperature; nothing can be colder.

EXERGY

In **1873 J. Willard Gibbs** invented the idea of available energy—called exergy (from the **Greek ex and ergon**, meaning "from work," by **Zoran Rant in 1956**). The **exergy** of a system is the **maximum useful work possible** during a process that brings the system into **equilibrium** with a **heat reservoir**.

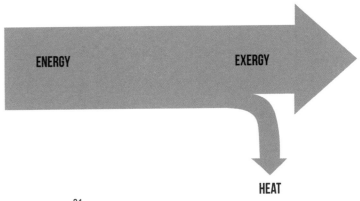

ENERGY

EXERGY

HEAT

TRUSS CONSTRUCTION

A truss is an assembly of straight members connected at joints, usually in triangles, so that the whole assembly acts as a single object. Each member is either in compression or in tension.

TYPES OF TRUSS

The **rafters and ceiling joist** form a simple truss, in which the **sloping members** are in **compression** and the **horizontal one** is in **tension**.

The **basic diamond frame** of a bicycle is a **planar truss**, made of **two triangles**. Such a truss has much **lower weight** than would the same beam or structure made from **solid material**.

A **space frame** truss is a **three-dimensional lattice** of triangles used, for example, in pylons carrying **high-tension electricity**.

The **king-post truss**, brought to England around **1614** by **Inigo Jones**, allowed **Christopher Wren** to **build houses** with large **flat ceilings**, because the **vertical** member is in tension, and holds up the **horizontal** member.

The **queen-post truss** is an **extended version** of the king-post truss.

In **1844 Caleb and Thomas Pratt** patented the **Pratt Truss**, which was much used in **building bridges**, in **wood**, then **iron**, and finally in **steel**.

In **1820** American engineer **Ithiel Town** patented the **Town truss**, widely used subsequently for **lattice bridges**. The many small, closely spaced, **diagonal elements** could be made of **planks** and **assembled** by unskilled labor.

VISCOSITY

A thick, glutinous, sticky fluid is said to be viscous. In practice viscosity is the amount of internal friction; the force-per-unit area resisting steady flow.

STOKES' LAW

Sir George Stokes investigated **viscosity** in the **1840s**. Stokes' Law allows the scientist to **calculate the viscosity of a liquid** by dropping a **ball bearing** down through a **tube full of liquid**, and noting its **terminal velocity**.

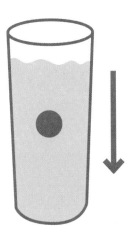

FRICTION

When a **liquid (or gas)** flows through a tube, it gets held up by friction at the **edges**, and successive layers move more quickly as you go **toward the center**. Viscosity is a **measure of the friction** between **successive layers**.

EXPLAINING VISCOSITY

Syrup is much **more viscous than water**. **Thixotropic liquids**, including tomato ketchup, become less viscous when **shaken or stirred**.

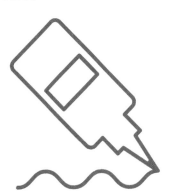

EFFECTS OF AIR AND WATER

The viscosity of air leads to the **force of drag on aircraft**; the viscosity of **water slows boats and ships**. Aircraft and ships have **streamlined shapes** designed to **minimize this drag**.

EFFECT OF TEMPERATURE

At temperatures close to **absolute zero**, the viscosity of **superfluids** such as **helium-3 and helium-4** drops to zero. If some is placed in a cup, a **thin film creeps up the sides**, over the rim, and down the outside.

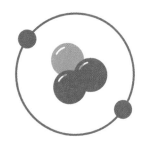
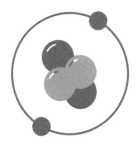

VULNERABILITY

Engineers can reduce vulnerability both by assessing and allowing for the risks before embarking on a project, and by providing ways of coping with disasters when they happen.

NATURAL DISASTERS

The vulnerability of a **person**, **system**, or a **building** is its inability to cope with the effects of a **hostile environment**. For example, a **building** on **low-lying land near a river** may be **flooded**.

In **earthquake** zones, many engineering structures may be vulnerable to violent **tremors** (see page 14).

COMMON VULNERABILITIES

Elderly people may have **difficulty** managing **stairs**.

In the United States, **car crashes** kill **35,000** people every year; anyone traveling in a car is vulnerable.

DANGEROUS PURSUITS

Some people deliberately make themselves vulnerable; for example **rock climbers**, **skiers**, and **scuba divers** all choose these activities for the **excitement of risk**.

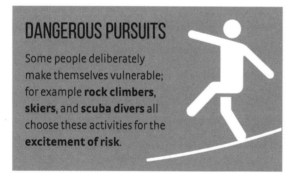

DANGEROUS SUBSTANCES

Emergency service personnel may need **protective clothing** in order to deal with **dangerous substances**.

MITIGATING VULNERABILITY

The task of the engineer is to **try to avoid** making structures and people vulnerable, by gauging the risks involved in their work (see page 28).

This might mean **designing bomb shelters** in the case of **war**...

... or better **helmets for cyclists**.

TOILETS

The toilet is an elegant engineering solution to the challenge of disposing of human effluent. Toilets were a feature of some of the earliest human civilizations, but the toilet as we know it emerged in the eighteenth century and has changed remarkably little since.

c.2500 BC First toilets, with wooden seats and sewage channels, installed in the Indus Valley cities of Mohenjo Daro and Harappa

c.2000 BC First flushing toilet installed in the Minoan palace of Knossos

c.100 BC Roman communal toilets, flushed with running water

c.AD 1200 Medieval castles feature garderobes

1596 Sir John Harrington designs for Elizabeth I a privy with an overhead cistern and lever-operated valve

1775 British watchmaker Alexander Cummings patents the S-bend, or "stink-trap"

1778 Flushing toilet patented by Joseph Bramah of Yorkshire, England

1885 British porcelain manufacturer Thomas Twyford and sanitary engineer J. G. Jennings together create the first one-piece, all-china toilet

TOILET PAPER

Before toilet paper, bums were cleaned with items ranging from **sponges on sticks** (the **ancient Romans**) to **wool** or **cotton** (rich **European merchants**) to **corn cobs** (**colonial Americans**). Paper was being used to clean backsides as long ago as sixth-century **China**, but paper specifically for use in the bathroom was invented by the American **Joseph Gayetty** in 1857. In the 1880s the **Scott brothers** began selling toilet paper on a roll.

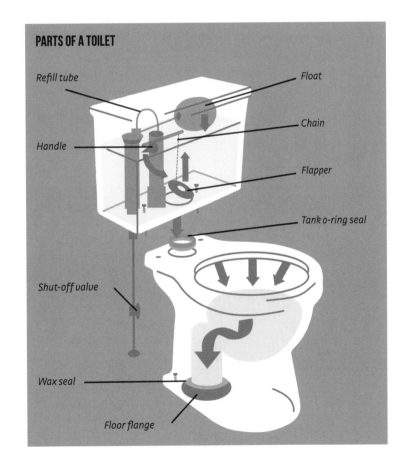

PARTS OF A TOILET

Refill tube

Float

Chain

Handle

Flapper

Tank o-ring seal

Shut-off valve

Wax seal

Floor flange

DAMS

A dam is a structure built across running water to retain water for various purposes, including for consumption, irrigation, as a power source, to improve navigability, to provide recreational possibilities, or to control flooding and flow rates.

PARTS OF A DAM

The elements of a dam are defined by its **upstream-downstream axis**. The **heel** of the dam is where the upstream face meets the foundations; the **toe** is where the downstream face meets the foundation. The water that is impounded (retained) at the upstream side is the **pool**; the water below the dam is the **tail**. The **freeboard** is the **clearance** between the **surface** of the pool and the **lowest point** where it could **overflow** the dam. A **spillway** is a channel for water to be discharged over or around the dam. **Abutments** are the **natural** or **artificial** parts of the walls of the valley against which the dam is constructed.

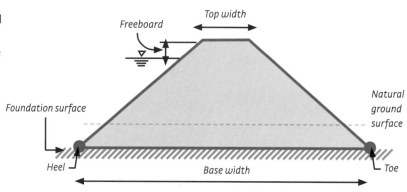

BEAVER DAMS

Humans are not the only animals to construct dams. The average beaver dam is about 6 ft. high and 5 ft. wide, but the biggest beaver dam ever spotted, on the southern edge of **Wood Buffalo National Park in Alberta, Canada,** is 2,788 ft. wide and probably took twenty years for the beavers to build.

TYPES OF DAM

A dam needs to be able to withstand the **forces exerted by the water it retains**. The main types of dam achieve this in different ways:

- **Arch**: Curving the dam in the horizontal plane produces an arch, which transfers the force to abutments on the side of the valley. These are commonly used in **narrow, steep-sided valleys**.
- **Buttress**: Buttresses on the downstream side help **hold up the dam**, allowing less material to be used in the sections between the buttresses.
- **Embankment**: These have a wide cross section and are

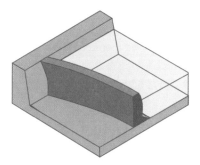

suited to **wider valleys**. They use **natural materials** such as **soil** and **crushed rock**, and usually feature an **impermeable central core**.
- **Gravity**: These rely on their **own mass** for **stability**, and are usually **triangular** in cross section.

THE BRONZE AGE

The Bronze Age is the name traditionally given to the stage of technological development that followed the Stone Age, when humans learned to work metals. It started around 4500 BC in the Middle East and gave way to the Iron Age as late as 500 BC in northern Europe.

- **c.6500 BC** First copper objects in eastern Anatolia

- **c.4500 BC** Start of the Chalcolithic (copper-stone) Age, a.k.a. the early Bronze Age

- **c.4000 BC** Metallurgists in Anatolia discover bronze by alloying copper and arsenic

- **c.3500 BC** Copper metallurgy widespread in Mesopotamia

- **c.3000 BC** Chalcolithic era in the Mediterranean, spreading to Europe; in Mesopotamia, tin replaces arsenic in bronze alloys

- **c.2500 BC** Bronze first appears in Britain

- **c.2000 BC** Tin deposits in Cornwall in use

- **c.1200 BC** Start of the Iron Age in the Middle East

- **c.1000 BC** Lead added to bronze

- **c.500 BC** End of the Bronze Age in Europe

- **15th century AD** Bronze Age in Mexico

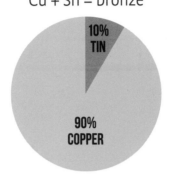

EARLY TIN-BRONZE

$$Cu + Sn = bronze$$

10% TIN

90% COPPER

COPPER AND BRONZE

Bronze is an alloy of copper, usually with **tin**. When **prehistoric peoples** developed the technology of **metallurgy**, they began with copper, which could be found in native (naturally occurring pure) form, and **alloyed** it with other elements to produce bronze, which is easier to work and produces stronger **tools** and **weapons**.

BIRTH OF CIVILIZATION

Control of bronze-making resources and technology helped drive the development of **city-states** and **civilizations**, so that the **Bronze Age** saw the **emergence of the first civilizations in Mesopotamia**, with **writing**, **accounting**, and **advanced mathematics**.

BRONZE AGE ADVANCES

Advances in technology and engineering during the Bronze Age included massive **irrigation projects**, construction of **pyramids** and other **monumental buildings**, the spread of **wheeled carts and chariots**, the invention of the **plough**, and eventually the development of **ironworking**.

BRIDGES

Bridges come in all shapes and sizes. As high-speed travel continues to shrink the world, engineers will no doubt build ever longer and grander bridges to cross stretches of water.

SIMPLE BRIDGES

For thousands of years people have **dropped logs or planks over streams** in order to get across.

On regular routes simple bridges were **made of stone**.

In South America 500 years ago, the **Inca people** made **rope bridges** to cross **canyons and gorges**, and renewed the ropes every year, using **ichu grass**. In some places this process is still maintained.

Meanwhile, major paths in developed countries had **stone-built bridges** to cross rivers.

TRANSPORT BRIDGES

As demand for transport grew, bridges had to be strong enough to **support horses and carts**, and **flat enough to carry railroad lines**.

Wooden railroad bridges were **cheaper and quicker** to build.

When **iron** became available in sufficient quantities, it became the **first choice** for bridge building. At **Coalbrookdale** in England, the iron bridge at what came to be called Ironbridge was built to show what was possible.

TUNNELS

A tunnel is a passage excavated underground. The earliest tunnels were dug as aqueducts, but they are also used for transport, storage, and carrying pipes and cables.

- **c.4000 BC** Qanats (aqueduct tunnels) constructed in Persia

- **6th century BC** Ancient Greek engineer **Eupalinos** constructs aqueduct tunnel through a mountain on the island of Samos

- **c.1st century BC** Romans develop military tunnels for undermining defenses

- **18th century AD** Boom in canal construction leads to advances in tunnel engineering

- **1880** First attempt to dig a tunnel under the English Channel

- **1991** Channel Tunnel completed

THE EUPALINOS TUNNEL

The Eupalinos tunnel is approximately 4,000 ft. long, with a diameter of 6 ft. and a ditch from 13 to 29.5 ft. deep. The maximum depth of the tunnel, beneath the summit of a mountain, is 558 ft. and it took eight to ten years to dig, teams starting from each end and meeting in the middle.

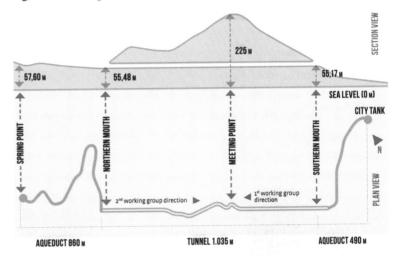

SECTION VIEW — 225 M — 57,60 M — 55,48 M — 55,17 M — SEA LEVEL (0 M) — CITY TANK — SPRING POINT — NORTHERN MOUTH — MEETING POINT — SOUTHERN MOUTH — N — PLAN VIEW — 2nd working group direction — 1st working group direction — AQUEDUCT 860 M — TUNNEL 1.035 M — AQUEDUCT 490 M

PARTS OF A TUNNEL

The openings of a tunnel are called the **portals**. The sides are called **walls**. The top half of a tunnel is called the **crown**; the bottom half is the **invert**. The point where the tunnel wall breaks from sloping inward to sloping outward is the **springline**. The excavated face of a tunnel is the **heading**.

TYPES OF TUNNEL

Basic tunnels are circular in cross section, and take the form of **two continuous arches joined together**. A **horseshoe tunnel** has a **flatter invert**. A **D-shaped tunnel** has a **flat invert**, **vertical walls**, and an **arched roof**.

SEIKAN TUNNEL

The **longest undersea tunnel** in the world (as of 2019) is the **Seikan Tunnel** in **Japan**. Connecting the Japanese islands of **Honshu** and **Hokkaido**, the tunnel is 33.46 miles long.

ARCHES

Arches are simple compression structures that engineers have used in buildings for thousands of years. They are curved structures that may be just an opening, like a doorway, or may bear the weight of a load above.

HOW DOES IT WORK?

The **keystone** pushes down and sideways on the stones below. The **point of the arch** is to **cross a gap by using compressive forces but nothing in tension**. The **compressive forces push the sides** of the arch outward; they are often supported by **abutments** on the outside.

ANCIENT ARCHES

People have been building arches for **four thousand years**, although many **early arches were underground**, where the surrounding soil supports the sides of the arch.

WHAT DOES IT MEAN?

The words *architect* and *architecture* mean "arch builder" and "arch building."

ROMAN STRUCTURES

Romans were enthusiastic arch builders, using **stone structures** to cover considerable heights and distances. One such was the **Pont du Gard aqueduct**; the top level carried water to a nearby city. They also included arches in buildings, and built **vaults and domes**, which are **three-dimensional arches**.

MEDIEVAL ARCHES

Most **Roman arches** were **round at the top**, but the builders of **medieval gothic** churches and cathedrals often used **pointed arches**, which generate **less sideways force at the bottom**.

TWENTIETH-CENTURY ARCHES

Swiss civil engineer **Robert Maillart** was a bridge designer who **specialized in reinforced concrete** arches. His **Salginatobel Bridge**, built 1929–30, is a **three-hinged bridge**. The hinges at both ends and in the middle allow for **thermal expansion and contraction** of the structure.

The **Bayonne Bridge**, designed by **Othmar Amman** and opened in 1931, connects Bayonne, New Jersey, with Staten Island, New York. It is one of the world's **longest steel arch bridges**.

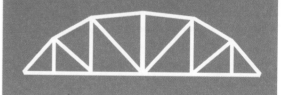

LIGHTHOUSES

A lighthouse is a structure that employs lights and other aids to help navigators, usually of ships at sea, and to warn of dangers such as rocks.

660 BC The poet Lesches describes a lighthouse at Sigeum (the present Cape Inchisari in Turkey)

c.300–280 BC Sostratus of Cnidus builds Pharos lighthouse in Alexandria

C. AD 50 Emperor Claudius builds a great lighthouse at Portus near Rome

c.150 First lighthouses in Western Europe built by the Romans at Dover and Boulogne

c.800 First lighthouse built out at sea, a Cordouan rock in the Gironde estuary

c.800 Record of navigational beacons at the entrance to the Persian Gulf

c.1200 The Mayans construct lighthouses in Central America

1562 There are 711 beacons along the Chinese coast between western Kuangtun and northern Chiangsu

1763 First catoptric system, which reflected oil light off parabolic mirrors

1800 Inland lighthouse at Pica, in Chile, to guide travelers across the Atacama Desert

1822 French physicist Augustin Fresnel perfects the dioptric lens and prism system

1902 Introduction of the parabolic reflector fitted with an electric arc light

1930s First automated lighthouses

TWO OUT OF SEVEN

Two of the **Seven Wonders of the Ancient World** were lighthouses: the **Pharos of Alexandria** and the **Colossus of Rhodes**.

THE PHAROS

The ancients claimed that the Pharos was 600 ft. high, but in reality it was probably around 150 ft.

FRESNEL LIGHT

Augustin-Jean Fresnel perfected an ingenious system that helped to capture and direct the maximum amount of light from the source. **Dioptric prisms** refract light so that it travels out horizontally, while, above and below, **catadioptric prisms** bend light rays almost perpendicularly, so that they also travel horizontally.

ARCHIMEDES

The Greek Archimedes was primarily a mathematician, but he was also an engineer of extraordinary skill and ingenuity.

 DATES C.287–212 BC

ARCHIMEDES' DISCOVERIES AND INVENTIONS

PI

He showed that the value of π **(pi) lies between 3.1408 and 3.1429**.

$$\frac{\text{circumference}}{\text{diameter}} = \pi$$

SPHERE INSIDE A CYLINDER

Using sheer logic, and without the equations that we have today, he showed that the **volume of a sphere inscribed in a cylinder is two thirds of the volume of the cylinder**, and that the surface areas are in the same ratio. He was so proud of this that it was **engraved on his tombstone**.

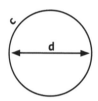

ARCHIMEDES' PRINCIPLE

This states that if a **body is immersed in water, it experiences an upthrust equal to the weight of the water displaced**. He famously jumped naked out of the public bath, **shouting "Eureka,"** when he discovered this.

ARCHIMEDES' SCREW

This screw acted to **remove bilge water** and stopped the huge ship *Syracusia* leaking from the hull.

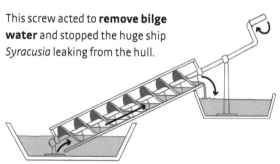

LEVERS AND PULLEYS

He worked out the **law of the lever and pulleys**, allowing sailors to lift heavy loads that were previously impossible.

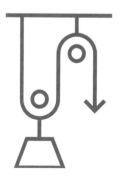

HEAT RAY

He invented the **heat ray** and **sank invading Roman ships**.

A GOOD CITIZEN
Many of the **inventions of Archimedes** were born of the need to help his **native city** of Syracuse, especially **protecting it from invasion**.

WATERWHEELS

A waterwheel is a device for converting the kinetic and/or potential energy of water into rotary motion, to power processes such as grinding or lifting.

WHEEL HISTORY

The first written reference to a waterwheel dates to 400 BC, but they were probably in use many centuries before this. Until the **Industrial Revolution**, water was the **primary nonanimal source of power** and waterwheels might be found on almost every river or stream.

NORSE, OVER, OR UNDER

There are three basic types of waterwheel. The simplest, oldest, and least efficient is the side-turning or **"Norse" wheel**. More efficient is the **undershot vertical wheel**, and most efficient is the **overshot vertical wheel**.

DOMESDAY WHEELS

The **Domesday Book**, a record of property in England compiled in 1086, records 5,624 waterwheels, although there were probably many more.

CAMS

Mills often needed **gears** to turn the motion of the waterwheel into useful work. An ingenious simple method to **convert rotary motion into linear motion** for processes such as **stamping** or **mashing** is to fit a **cam** to the wheel and another to a **rod** or **beam**.

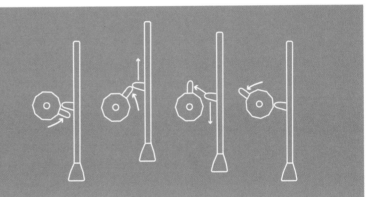

OVERSHOT WATERWHEELS

Undershot wheels are **less efficient** and will stop working altogether if the water level falls. If the drop of the watercourse is high enough, or works can be done to create an artificial drop, the wheel can be set up so that the flow of water falls onto the wheel from above, adding potential energy from **gravity** to the **force of the flow**. Different types of overshot wheel vary in efficiency.

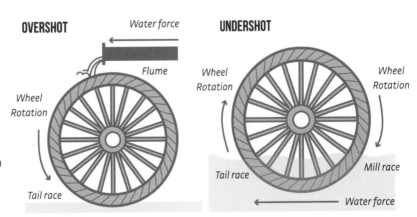

OVERSHOT — Water force — Flume — Wheel Rotation — Tail race

UNDERSHOT — Wheel Rotation — Wheel Rotation — Tail race — Mill race — Water force

WATER SUPPLY AND SEWERS

The Romans achieved many engineering landmarks, but among their most impressive achievements were their hydraulic structures, including aqueducts, baths, and sewers.

AQUEDUCTS

To bring water from sources to the city, the Romans relied on **gravity** and **clever engineering**. Contrary to popular belief, aqueducts rarely crossed bridges; most were in **tunnels** or **covered channels** that hugged **contour lines**. To cross gaps, engineers could use **bridges** or **siphons**.

21 million cubic feet: the estimated daily water supply to Rome.

SEWERS OR DRAINS?

Most **ancient runoff channels** characterized today as **sewers** were actually for draining surface water from **marshes**, **floods**, and **storms**. The **Cloaca Maxima** ("Greatest Sewer"), a huge tunnel in Rome, was used for these purposes. Only later were some public and private **latrines** connected to it, and until the nineteenth century, sewers specifically for **human waste** were very rare.

CLOACA MAXIMA

Started in the sixth century BC as an open channel for **draining marshland**, the Cloaca Maxima was enclosed in the third century BC and converted into a giant, **barrel-arched vault**. It remained in service for 2,400 years.

ROMAN LATRINES

Domestic latrines emptied into **pans**, **buckets**, or **cesspits**. **Public latrines** were **long benches** over **water channels** to carry away the waste, with another channel at the user's feet for **washing** and **rinsing**.

BATHS OF CARACALLA

One of the greatest buildings in ancient Rome, the **Baths of Caracalla** complex covered 13 hectares, and the main building was 750 ft. long, 380 ft. wide, and, it is estimated, 125 ft. high. It could hold 1,600 bathers at a time. The complex included **two libraries**, **gymnasia**, and **shopping malls**. It was **heated** with a **hypocaust** and **hot water** from a dedicated **aqueduct**.

THE GREAT WALL OF CHINA

One of the great feats of engineering in human history, the Great Wall is actually a collection of stretches of wall scattered across China, taking many different forms.

- **5th–3rd century BC** Warring states—minor walls constructed between kingdoms, along with northeastern section
- **221–206 BC** Qin Dynasty—rebuilds northeastern section, extends west
- **206 BC–AD 220** Han Dynasty—rebuilds northeastern section, extends far to the north and east, deep into the Gobi Desert
- **386–584** Northern Wei Dynasty—central section west of Beijing
- **1066–1234** Liao and Jin Dynasties—extra sections to the far north
- **1368–1644** Ming Dynasty—rebuilds the wall around Beijing, northeast to the Korean border, and scattered sections in the west

HORSES FOR COURSES

While the classic image of the Great Wall is a **brick and stone** structure, this only accounts for parts. In the **desert regions**, **sand and gravel** would be used to fill between **screens of willow and reed**. In areas without much stone, **earth** and **soil** would be rammed down to create vast **berms**.

CLASSIC CONSTRUCTION

The classic **Ming Dynasty** wall was built by creating **parallel walls** of **stone slabs** or **kiln-fired bricks**, cemented with a **mortar of lime** and, sometimes, **glutinous rice**. The space between the walls was filled with a **bed of locally quarried rock**, on top of which **earth** and **rubble** was tamped. Once the top of the wall had been reached, a **layer of brick and stone** paving was laid.

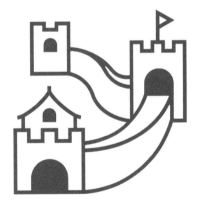

TEN-THOUSAND LI

Traditionally, the Great Wall was known as the "**Ten-Thousand Li Wall**." A li is roughly 500 yards, and the length of the original wall constructed by the first emperor **Shi Huangdi** is said to have been 3,100 miles. The Chinese government says that the **total length of all the fragments of the Wall** built by many different dynasties is 13,171 miles. The Great Wall that tourists visit today is the **Ming Dynasty wall**, said originally to have been over 5,500 miles long.

The **Ming Wall** stands up to 26 ft. high, averaging 15–30 ft. in thickness at the base and sloping to

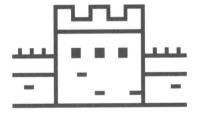

12 ft. thick at the top. It was **fortified** by 25,000 **towers** and 15,000 **outposts**.

DOMES

A dome is a form of vault, an arch that is deeper than its span. The dome may be circular, elliptical, or polygonal in plan, and bulbous, segmental, semicircular, or pointed in vertical sections. The Romans discovered a new material that allowed them to build enormous domes.

CIVIL ENGINEERING

ROUND OR SQUARE

Domes vary according to the plan of the space they sit atop. If the **supporting structure** has the **same plan** as the dome, it is known as a **drum** and the whole structure is a **rotunda**. But if the plan of the supporting structure is **square or rectangular**, intervening elements make possible the **transition from square to circle**; these include **lunettes**, **pendentives**, and **squinches**.

TYPES OF DOME

Domes come in a weird and wonderful variety of names, including **umbrella**, **melon**, **sail**, **calotte**, **cloister**, **parachute**, **pumpkin**, and **Pantheon dome**. The latter type, named after the mighty **Roman Pantheon**, is a low dome that is **coffered** on the inside and often **stepped** on the exterior.

THE PANTHEON

A great **temple** to all the **Roman gods**, this building in Rome was started by **Agrippa** in 27 BC but not completed until the reign of **Hadrian** (C.AD 120). By using **concrete**, the Romans were able to build a **huge dome sitting on top of a drum** to give a **rotunda** 142 ft. in both diameter and height, lit from a central **oculus** (circular opening) 30 ft. wide.

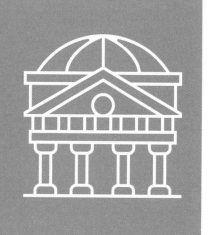

BRUNELLESCHI'S DUOMO

The **dome of the cathedral in Florence** was a revolutionary design by **Filippo Brunelleschi**, who (starting in 1420) built a **double-skinned** dome complete with innovations to bind the dome together, including **iron chains**; **tension rings** of **stone**, **iron**, and **wood to counteract "hoop-stress"** (outward, spreading pressure); and **spiraling** courses of **herringbone brickwork**. Even his **self-designed pulleys and cranes** were innovative.

WINDMILLS

A structure with sails to catch the wind and drive a mechanism to do work.
Most windmills were used to grind grain or pump water. They were widely
used until the advent of the internal combustion engine.

1st century AD Heron
of Alexandria describes
a wind-powered machine

644 Earliest known reference
to a windmill in Persia

915 Horizontal windmills
in Persia

c.1200 Genghis Khan
brings Persian windmill
wrights back to China

c.1200 Post windmills in Europe

1420 Fixed tower mills (a.k.a.
smock mills) in Europe

1745 Edmund Lee in England
invents the automatic fantail

1772 Andrew Meikle in Scotland
invents the spring sail

1789 Stephen Hooper in
England develops remote-
controlled roller blind sails

1807 Sir William
Cubitt invents
"patent
sail," which
combines
Meikle and
Hooper innovations

1854 Daniel Halladay in the
United States invents the
annular-sailed wind pump

1890 First electricity-generating
windmill built in Denmark

TYPES OF WINDMILL

Three main types of windmill developed over time:

Horizontal mills
Similar to a **"Norse" water mill**, the sails are mounted in the **horizontal plane** on a **vertical axis**, so that **no gearing is needed** to drive a mill.

Tower or smock mills
The **sails** and **axle** are contained in a **rotating cap** that sits on a **fixed tower**, in which the rest of the machinery and structures are contained. Can be much larger.

Post mills
The **sails** are in the **vertical plane** attached to a **horizontal axis**, with **gearing used to drive rotation on other planes**. The sails, gears, and other machinery are all contained in a structure that sits on a sturdy post, so that the whole **can rotate to face into the wind**.

CIVIL ENGINEERING

IRON

Iron has tremendous strength for its weight, and is easy to work, widely available, and cheap—if the technology exists to smelt it properly. Iron was a transformational material for engineering.

CIVIL ENGINEERING

- **c.3000 BC** Meteoric iron in use
- **c.1200 BC** Start of the Iron Age, with smelting of iron ores
- **c.500 BC** Etruscan iron-smelting center Populonia producing 1,600–2,000 tons of iron per year
- **c.500 BC** Blast furnaces in China producing pig iron
- **c.AD 1** The Haya people of East Africa independently invent blast furnaces
- **c.1200** Blast furnaces in Western Europe
- **1708** Abraham Darby in England introduces coke-powered blast furnace
- **1754** First iron rolling mill for industrial production of wrought iron
- **1760** British annual production of pig iron c.2,500 tons
- **1779** First cast-iron bridge, spanning the Severn at Coalbrookdale, made by Abraham Darby, using cast-iron beams
- **1805** British annual production of pig iron c.350,000 tons
- **1850–51** Joseph Paxton builds the iron and glass Crystal Palace in London

BLOOMS

Iron has a very **high melting point** (1,540°C; 2,804°F), and before **blast furnace technology** made such temperatures achievable, **metallurgists** had to obtain it from blooms—mixed masses of **iron**, **charcoal**, and **slag** produced by heating, out of which pieces of purer iron would be hammered and worked together to produce **wrought iron**.

PIGS

Blast furnaces, where air is forced in under pressure, can achieve **higher temperatures**, so that **molten iron** can be run out of the furnace and left to set in molds that resemble piglets suckling on a sow. The resulting **ingots**, known as pigs (hence "pig iron"), are then used to produce **wrought iron** or **steel**.

CAST VS WROUGHT

Cast iron (molten iron poured into forms and left to cool) is **strong in compression** but **weak in tension**. In buildings, it is used for **columns** and **railings**. **Wrought iron** (iron beaten to strengthen it) is less **brittle**.

IRON BUILDINGS

The **Ironbridge** at **Coalbrookdale** in England was the first large structure built with iron. It was followed by more bridges and then **iron-framed factories and warehouses**, such as **William Strutt's North Mill at Belper** in 1804. **Cast-iron columns** were used to carry **beams** to **support brickwork**. **Victorian iron and glass structures** reached their high point with the **Crystal Palace** in 1851. **Iron girders and rivets** made it possible to construct **metal frames** with **light cladding**, leading to ever taller buildings.

STEAM ENGINES

The defining technology of the Industrial Revolution, steam power would revolutionize the world, but it needed some brilliant engineering to get it to work.

- **1st century BC** Hero of Alexandria describes an aeolipile or steam engine

- **1644** Evangelista Torricelli demonstrates that the atmosphere has weight and proves existence of vacuum

- **1657** Otto von Guericke uses vacuum and atmospheric pressure to create tremendous force

- **1679** Denis Papin produces partial vacuum by condensing steam and creates a single-stroke piston

- **1698** Thomas Savery patent for a pump that worked by condensing steam in a cylinder to draw up water with the consequent suction

- **1712** Thomas Newcomen demonstrates his steam engine pump for extracting water from mines

- **1769** James Watt patent for separate condensing chamber

- **1775** Watt and Matthew Boulton produce efficient steam engines

- **1781** Watt redesigns his pure steam engines to produce rotary motion for a wider range of industrial application

NEWCOMEN OPENS THE DOOR

Newcomen's **steam pump** could be powered with **discarded coal** while at the same time making viable a **massive expansion of coal mining**, fueling the **Industrial Revolution**.

PUMP ⟶ COAL ⟶ FACTORIES

WATT'S CONDENSER

Newcomen engines inefficiently had **steam condensing in the cylinder**, which thus had to be cooled and heated with each stroke. Watt realized that steam is **"elastic,"** and so could create **pressure** (via a partial vacuum) in one vessel (which could be kept hot) while **condensing** in another (which could be kept cold).

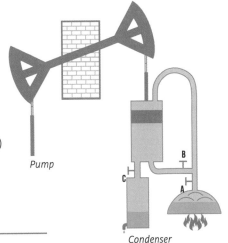

Pump

Condenser

ATMOSPHERIC VS PURE STEAM

Newcomen's engine, and **Watt's early models**, were **atmospheric engines**: they relied on the **weight of the atmosphere** to **drive the down stroke of the piston**. Watt later **enclosed the piston** and used the **expansion of steam** to drive **both strokes of the piston**, massively **increasing its efficiency**.

FACTORIES

A factory is a system for manufacturing; normally, it is understood to be a building or set of buildings, but in the engineering sense it can be viewed as a way of combining and regulating components into a system.

CIVIL ENGINEERING

BEFORE FACTORIES

Before the **Industrial Revolution**, a factory was a **trading depot** that served as the **base for a trading agent or factor**. **Industrial processes**, on the other hand, tended to be small-scale and distributed: **cottage industries**. But there were always **exceptions**. At **Laurium**, in **Greece**, for instance, there were *ergastéria*: ancient factories for **processing and smelting ore**.

ARKWRIGHT'S MILLS

Richard Arkwright was an English barber who developed **machines for textile manufacture** and put together a number of **mechanical and process innovations** under one roof at a **mill** on the banks of the **River Derwent** in Derbyshire, in 1771, to create the **first industrial factory**. Although usually described as an **innovative entrepreneur**, he is perhaps more accurately described as a **systems engineer**.

SCIENTIFIC MANAGEMENT

American innovations in factory systems, especially **intense mechanization**, led the British to talk about the "**American System of Manufactures**." These innovations in systems engineering would reach new heights with **Frederick Taylor's Scientific Management principles**, which advocated a clear **division of labor**, **centralized control**, **rational management**, and "**stopwatch analysis**" of processes.

AMERICAN EXPERIMENTS

British factory innovations were **imported to the United States** by engineer **Samuel Slater**, who built the **first textile factory in the United States** on **Rhode Island** in **1790**. Slater went on to develop "**corporate villages**" in which the whole community contributed to the **manufacture of cloth**, extending the definition of a factory. In **Massachusetts** from **1814**, a group of industrialists led by **Francis Cabot Lowell** developed an **alternative model**, with **large, fully integrated, highly mechanized factories**.

SUSPENSION BRIDGES

*The classic example of a tensile structure, made up of elements acting in tension,
the form of a suspension bridge is determined by its own weight.*

ROPES AND CHAINS

Suspension bridges made of **vines** and **ropes** were probably some of the **earliest engineered structures built by humans**. **Iron-chain** suspension bridges were built in **South Asia** in the **Middle Ages**.

JACOB'S CREEK

The **first suspension bridge in the Western world** was designed by an American judge, **James Finley**, to cross **Jacob's Creek in Pennsylvania**. He copied **Asian models** and designed a bridge with **decking** that hung on **vertical suspenders** dropped from an **iron-chain cable**.

MENAI BRIDGE

In 1826, **Thomas Telford** completed the **Menai Suspension Bridge**, which hangs from **iron chains**, joining **Wales** to the island of **Anglesey**. The **Royal Navy** insisted that there must be **clearance** of 100 ft. above **high tide**.

JOHN A. ROEBLING SUSPENSION BRIDGE

Roebling was a **pioneer** of suspension bridge design. His 1,057 ft. bridge linking **Cincinnati** with **Covington** was the **longest in the world when it was opened in 1866**.

GEORGE WASHINGTON BRIDGE

The George Washington Bridge, designed by chief civil engineer **Othmar Ammann**, is 4,760 ft. long and is suspended by 106,000 **wires**, woven into **four main cables**, and carries **fourteen lanes of traffic** to and from **New York City**.

DANYANG–KUNSHAN BRIDGE

At 102 miles, the **longest bridge in the world** is the **Danyang–Kunshan Grand Bridge in China**, part of the **Beijing–Shanghai High-Speed Railway**.

CEMENT

Cement is a vital component of mortar, a binding substance that is used to join together other materials, such as bricks, or mixed with aggregate to produce concrete.

GYPSUM

The ancient **Egyptians** made cement using gypsum, a **calcium-** and **sulphur**-rich rock. **Calcining** the gypsum (heating it to **drive off most of the water** chemically bound to the calcium and sulphur) and crushing it created a **powder** that could be mixed with water to create a **cementitious (sticky) mortar**.

ROMAN CEMENT

The Romans discovered a cement they called **pozzolan**: a kind of **volcanic ash** similar to powdered brick, originally from **Pozzuoli near Vesuvius**, which could also be made by using **burned clay and brick dust**. Their cement **could even be used underwater**.

PORTLAND CEMENT

The **secret** of **waterproof cement** was **lost until the late Middle Ages**. Modern cement traces its origins to **Joseph Aspidin**'s 1824 patent for **Portland cement**, which he named because it made concrete that resembled **Portland stone**. Aspidin's cement was made by **heating powdered clay with limestone** to **calcine** it. In 1845 **Isaac Johnson** used **higher temperatures** to create the **first modern Portland cement**.

HYDRAULIC POWER

Portland cement will **set even under water**, as dramatically demonstrated when a **ship** carrying barrels of Aspidin's cement **sank off the coast of England** and the **set barrels** were recovered and **used to build a bar in Sheerness**, in **Kent**.

TUNNELING SHIELD

Digging a tunnel beneath a river was one of the great engineering challenges of the nineteenth century. A initial attempt to dig under the Thames in London had met with tragic failure, but then engineer Marc Brunel (1769–1849) saw a worm that gave him an idea.

DANGEROUS GROUND

In 1807, **Cornish steam engine pioneer Richard Trevithick** (1771–1833) tried to dig a **tunnel** under the **Thames**, but discovered that it was an **especially difficult task**. A tunnel would have to pass through **waterlogged silt** and **quicksand**, under **tremendous pressure** from the **water** above, especially when the tidal river ran high. Trevithick's tunnel continually **flooded** with filthy water and eventually he **abandoned the attempt**.

TUNNEL WORM

French émigré engineer **Marc Brunel** saw a **rotting ship's timber** and examined a **teredo** or **shipworm** (actually a mollusc) with a magnifying glass. He saw that the creature **burrows** by cutting the face of the tunnel with **rasping jaws** and **processing the wood pulp into hard tunnel lining** as it goes. This inspired his **tunneling "shield."**

THE SHIELD

Brunel's brilliant idea was to construct a **grid or frame of metal**, protected by a **metal roof** and **divided into compartments, each sealed off at the front** with planks called **poling boards** to hold back **wet earth**, and each of which could be moved forward on **jacks**. Each compartment housed a **miner** who would remove a board at a time, **dig a box-shaped cavity** a few inches deep, and **push the board forward** to the **new face**. Once the whole cell face had been dug out, the cell would be **jacked forward** to press up against the new face, and the **whole shield also inched forward** by jack. **Masons** behind the rig quickly **lined the tunnel with bricks**.

SLOW WORK

In 1825 Brunel started work, but although the shield worked, the project remained **dangerous** and was **abandoned for seven years** when a **flood killed six miners**. It was **eventually completed in 1841**.

CONCRETE

Concrete is a type of artificial rock that can be molded and shaped, and used to join other building materials together.

MAKING CONCRETE

Concrete is made by **mixing** bits of hard material (known as **aggregate**, which is **normally crushed stone**) with a **mortar**, made from a mixture of **sand**, **water**, and **cement** (a **binding agent**, which today is normally **Portland cement**).

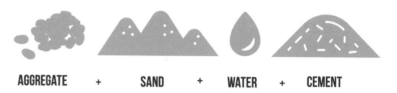

AGGREGATE + SAND + WATER + CEMENT

ROMAN CONCRETE BUILDINGS

Once they had learned how to make **slow-drying cement**, the Romans were able to build amazing concrete structures—including **vast domes** like the **Pantheon** and the **Domus Aurea**—and to create huge indoor spaces.

ROMAN CONCRETE

The Romans were the first to master the possibilities of concrete, using a type they called **opus caementicium**, made of **stones mortared with lime and pozzolan**.

REINFORCEMENTS

The first experiments with reinforcing concrete were made in the early nineteenth century. In 1832 **John Loudon**, a pioneer of **Victorian greenhouse architecture**, recorded the use of **iron bars** for **reinforced concrete floors**; others used **wire mesh**. Today, reinforced concrete is made with **steel rods**, so that it can withstand **enormous compressive and tensile loads**, with the added benefit of being very **fire resistant**.

TENSION AND COMPRESSION

Concrete is **strong in compression** but **weak in tension**. If concrete is to be used for components such as **beams** it must be **reinforced**.

Compression

Tension

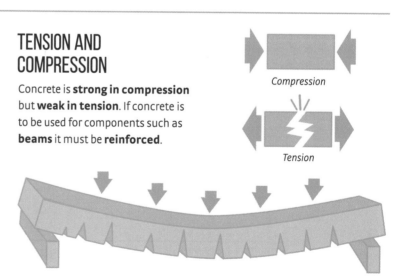

Tensile forces pull apart the bottom of this concrete slab when it bends

BESSEMER'S CONVERTER

Steel became perhaps the most important engineering material of the late nineteenth and twentieth centuries, but only after an engineer worked out how to make it reliably and cheaply.

CARBON CONTENT

Steel is an **alloy of iron and carbon** (0.1–1 percent by weight), and sometimes other elements such as **manganese**. Adding **carbon** to iron is called **carburization**. **Ancient iron smiths** could achieve **surface carburization** of iron by heating it with **charcoal**, but they **could do little to control or scale the process**.

PERFECT BALANCE

Carbon changes the **crystalline structure** adopted by the **iron atoms**, making it **harder** and increasing its **tensile strength**, but also more **brittle** (a.k.a. **less ductile**). The more carbon, the stronger and more brittle the steel becomes. Engineers need a **balance between tensile strength and ductility** and typically use steel with 0.15–0.4 percent carbon.

BESSEMER

Henry Bessemer (1813–98) was an engineer who had invented a new type of **gun** and wanted to be able to **cast it in steel**. He dreamt up a **large crucible** into which **molten iron** could be poured, and through which a **jet of air** would be **blasted** to **burn away the impurities** that made iron too brittle. This was his **converter**.

IMPURITIES

In the mid-nineteenth century, steel production was **limited** and **expensive**, so that its use was limited to small items such as **cutlery**. The alternative material, **wrought iron**, was also expensive as it had to be laboriously refined from **cast (pig) iron**, which contained **impurities that made it brittle**.

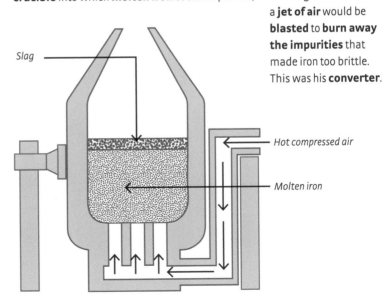

Slag

Hot compressed air

Molten iron

SIR JOSEPH BAZALGETTE

Not only did Bazalgette solve London's sewage problem—and his sewers are still in use today—but he was instrumental in eliminating cholera from the capital.

March 28, 1819 Born

1840 Population of London grows to 2.5 million; sewage from overflowing privies trickles through streets to the river

1849 Bazalgette appointed Assistant Surveyor to Metropolitan Commission of Sewers

1853–54 Cholera epidemic in London kills 10,000

1856 Metropolitan Board of Works (MBW) established with Bazalgette as Engineer

1858 Overflowing sewers and a hot summer make London unbearable. Michael Faraday wrote to *The Times:* "The smell was very bad . . . the whole river was for the time a real sewer."

1858 In the Houses of Parliament, MPs debate the Great Stink, and give MBW £3 million to build sewers

1874 Knighted

March 15, 1891 Died

BAZALGETTE'S SEWERS

Most of his early sewers were **built of brick**, but Bazalgette was one of the first engineers to develop the **use of concrete** as a structural material.

Pumping stations at **Abbey Mills**, north of the river Thames, and at **Crossness** to the south, raised the sewage enough to allow it to flow downhill to the outfalls further east. Abbey Mills is described as a "**Cathedral of Sewage.**"

On the sloping muddy banks of the river, Bazalgette built **Victoria**, **Albert**, and **Chelsea Embankments**, containing main sewers and underground railroad tunnels.

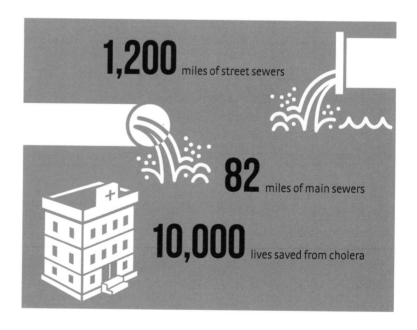

1,200 miles of street sewers

82 miles of main sewers

10,000 lives saved from cholera

SKYSCRAPERS

Very tall, multistory buildings became possible only when engineers developed the right construction materials and techniques and overcame major challenges.

1868–70 Equitable Life Assurance Building, NYC: **131 feet**

1883–5 Home Insurance Building, Chicago: **138 feet**

1887–8 Tacoma Building, Chicago: **164 feet**

1911–13 Woolworth Building, NYC: **791 feet**

1930–2 Empire State Building, NYC: **1,247 feet** (roof)

1969–70 John Hancock Center, Chicago: **1,129 feet** (roof)

1964–74 World Trade Center, NYC: **1,368 feet**

1972–4 Sears Tower, Chicago: **1,453 feet** (roof)

1991–7 Petronas Twin Towers, Kuala Lumpur, Malaysia: **1,482 feet**

2010 Burj Khalifa, Dubai: **2,716 feet** (roof)

THE SKY IS THE LIMIT?

The height of buildings was limited by a number of factors: **sufficiently strong materials** to bear **enormous loads**; **structural designs** that could cope with the loads; the **cost** of enormous quantities of **building materials**; and the number of **stairs** users would willingly climb.

INDUSTRIAL CONSTRUCTION

The strategies developed to build **bridges** and later **factories**, **prioritizing frames over walls**, would solve many of these problems. **Industrial construction methods** involved putting together **standardized components** in easily and rapidly **repeatable steps**, so that enormous structures could be **quickly and cheaply assembled**.

THE PRICE IS RIGHT

Although **designs** can be achieved to build skyscrapers with materials like **wood** and **brick**, the amount of such materials needed would be prohibitively **expensive**. **Iron** and especially **steel** are **strong enough** that **relatively light and cheap components** can be put together to create **frames** that **can bear huge loads**. **Off-site prefabrication** of components such as **concrete panels** and **steel beams**, in **standard sizes**, allowed cheap **mass manufacture**.

METAL FRAMES

Walls were traditionally the **load-bearing elements** of a structure, but **skyscraper design** turns this on its head. The load-bearing elements are the **metal frames** or **skeletons** of the building, off which walls hang like curtains—hence "**curtain-walls**."

GOING UP

It is not practical to expect people to walk up and down more than a few stories, so **elevators** were needed before skyscrapers could become viable. **The first office building to include elevators** was the **Equitable Life Assurance Building**, opened in **New York City** in **1870**; not coincidentally, it is **widely considered to be the first skyscraper**.

THE EIFFEL TOWER

Conceived as part of plans to celebrate the World's Fair of 1889 in Paris a century after the Revolution, Gustav Eiffel's great tower was a marvel of engineering in concept and execution.

EIFFEL THE ENGINEER

Though best known for his Tower, Eiffel also built many important **bridges** and other structures, including the **internal iron skeleton** of the **Statue of Liberty**.

PYLON

Eiffel and his engineers **Maurice Koechlin** and **Emile Nouguier** conceived of a **pylon-like structure** with four **columns of latticework girders** that would come together at the top. The design was inspired by the **Latting Observatory** in **New York**.

RIVETING STUFF

The metal pieces of the Tower are fixed together with **rivets**. These are like **bolts**, but before they are inserted, they are **heated** so that they **expand** and **soften**. Once in place they are **beaten** at each end **to flatten them out**, and as they **cool** they **contract** and **pull the pieces together**. **Two and a half million** rivets were used.

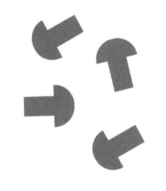

ORIGINAL VISION

The original design for the Tower had **architectural flourishes** such as **monumental arches, glass-walled halls** on each level, and a **bulb-shaped top**.

PRECISION ENGINEERING

At Eiffel's **workshop** on the outskirts of Paris, the **18,038 metallic parts** of the Tower were **designed to an accuracy of a tenth of a millimeter**, and then **prefabricated** into **five-meter-long elements**. These were shipped to the site, where 150 to 300 workers were **assembling the Tower**.

PUDDLING IRON

Eiffel was an expert at working with **puddling iron**, a **type of wrought iron** with **similar properties and benefits to steel**.

QUICK WORK

Ground was broken on January 28, 1887, and the Tower was **finished** on March 31, 1889, so that it took just **two years, two months and five days to build**.

IN NUMBERS

- Height: 1,122 ft.
- Storys equivalent: 84
- Length of sides at base: 410 ft.
- Weight: 7,300 tons
- Paint: 60 tons
- Elevators: 5

CHANNEL TUNNEL

The longest undersea tunnel in the world, the Channel Tunnel that runs between England and France, beneath the English Channel, is the realization of a dream more than two hundred years old.

1802 French engineer Albert Mathieu proposes a tunnel under the Channel, with an artificial island halfway across to allow horses to rest

1872 The coming of the railroads makes the project practical; the English Channel Company is formed

1881 The rival South Eastern Company is formed

1882 Digging begins at Dover

1883 Digging is abandoned

1966 The leaders of France and Britain pledge to construct a transport link between the two countries

1988 Digging begins from either end

1990 Tunnels link up

1991 Tunneling is completed

1994 Official opening by President Mitterrand and Queen Elizabeth II in May; the first travelers pass through in November

THREE IN ONE

There are actually **three tunnels**—two large railroad tunnels (25 ft. wide—high enough to take a double-decker bus) and a smaller **service tunnel** between them.

FOLLOWING THE CHALK

In order to stay within the **strata of stable chalk rock** that passes under the Channel, the Tunnel **curves up and down and left and right** in places. The **average depth below the sea bed** is 148 ft.

TRAFFIC JAMMED

The tunnel is used by **400 trains** a day, carrying, on average, **50,000 passengers**, **6,000 cars**, and **54,000 tons of freight**.

DIG FOR VICTORY

In total, eleven **tunnel-boring machines** were used: five on the French side and six on the British side. **Work on the French side was slower because the ground was wetter**. To **line the tunnel**, two systems were used: **precast concrete rings** and **cast-iron segments bolted together**.

TIME SAVER

The UK and France are about twenty-one miles apart at their **nearest point**. However, the three tunnels are thirty-five miles long, as they **run to terminals inland at Folkestone and Calais**. Before the Tunnel, it **took around six or seven hours by rail and ferry from London to Paris**. It **now takes two and a half hours by train**.

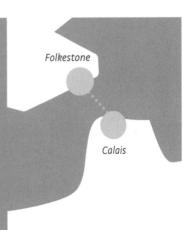

Folkestone

Calais

SAVING THE LEANING TOWER

The medieval bell tower of Pisa cathedral is famous for its lean, but by 1990 it was in danger of falling over. Engineers were called in to rescue the iconic building.

THE 184-FOOT-HIGH TOWER LEANS ROUGHLY 17 FEET FROM THE PERPENDICULAR.

SHIFTING GROUND

Work on the tower started in 1173, but did not finish until 1372! By the time the workers had built the second floor, **five years after starting**, **the tower had already begun to lean**, thanks to its **shallow foundations**, just ten foot deep, built on **unstable subsoil**.

COMPENSATING

As work on the tower continued, its builders sought to **compensate for the lean** by **building the upper floors** with the walls **on the north side** (away from the lean) **slightly shorter than the ones on the south side**, so that the tower **curves as well as leans**.

RESCUE SQUAD

In 1990 there were **fears that the tower might collapse**, and the Italian government appointed a committee led by **British engineer John Burland**. The team devised a **simple fix**: a few truck-loads of **soil** were **excavated from beneath the north side** of the foundations, while **steel cables** tied around the lower parts of the tower **stabilized it**. The **weight of the tower compressed the now less dense subsoil** on the north and by 2001 the **tower had straightened** by fifteen inches.

STILL GOING

By 2013 the tower had **straightened another inch**, and in the future it is expected to do so **a tiny bit more**, before **leaning again**—but **very slowly**, so it **should be safe for hundreds of years**.

THE WHEEL

The wheel was a transformative technology that emerged in western Asia by at least the early fourth millennium BC. Engineering led to steady improvements.

EVOLUTION OF THE WHEEL

The wheel evolved from **captive rollers**: **circular logs** placed beneath **sledges** to **reduce friction**, and held in place by **wooden dowels** or **pegs**. The rollers became **axles** with **solid wooden disks on either end**; the disk was fixed to the axle and the **whole assembly rotated**.

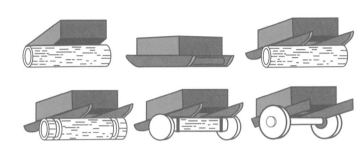

LINCHPINS AND NAVES

From the third millennium BC, it became standard for the **axle to be fixed to the carriage** with the wheel slotted over the end (held in place with a **linchpin**) so that it **rotated around the axle**. The part of the wheel through which the axle passes is called the **nave**.

LIGHTENING THE LOAD

Solid wheels are **heavy**. The first step in making them **lighter** was to **cut out sections** (known as making **lunate perforations**). Later, an **entirely new type of wheel** was developed, with a **bentwood rim** connected to the **nave** by **spokes**.

RANGE BOOSTER

On foot, a **person can travel** up to around thirty miles a day—far fewer if **burdened**. A **packhorse** with a **heavy load** can cover fewer than sixteen miles in an eight-hour day. The same horse **can transport twice as much**, three miles further each day, **if the load is mounted on a wheeled cart**.

THE MYSTERY OF THE MISSING WHEELS

The concept of the wheel was known to **pre-Columbian Americans** (in toys), but for unknown reasons they **never developed wheeled vehicles**.

CANALS AND LOCKS

*These are artificial or modified natural channels for inland water transport
and/or the control of water for irrigation or drainage.*

TRANSPORT

- **c.5000 BC** Irrigation canals in Mesopotamia

- **c.515 BC** Persian emperor Darius I constructs a canal linking the Nile to the Red Sea

- **c.610** Grand Canal in China links the Yangtze and Yellow rivers

- **c.950** Chinese engineer Chiao Wei-yo invents a two-level lock for canals

- **c.1500** First modern lock gates on a canal in Milan, probably designed by Leonardo da Vinci

- **1681** Canal du Midi is completed in France, including a 525-ft. tunnel

- **1825** Work begins on the 363-mile Erie Canal to link the Hudson River to Lake Erie

- **1869** Opening of the Suez Canal linking the Mediterranean to Red Sea

- **1914** Work finally completed on the Panama Canal

- **1959** Opening of the St Lawrence Seaway linking the Great Lakes to the Atlantic

POUNDS AND LEVELS

The **stretch of water** contained or "**impounded**" between **two locks** is known as a **pound**, **reach**, or level. **Summit pounds** are the stretches of level water at the **highest point on a canal**, and they need special arrangements to ensure that the water level remains **topped up**. **Sump pounds** are the **lowest stretches**. A **lock pound** is the relatively short **stretch between two locks**.

POUND LOCKS

Introduced in Europe around the fourteenth century, the pound lock has a chamber between two gates. It functions like an air lock: if the downhill gate is open, the water in the chamber is at the level of the downhill pound. Once the boat is in the chamber the gate is closed and water is let in from uphill, until the water level in the chamber is raised to the uphill pound level and the uphill gate can be opened.

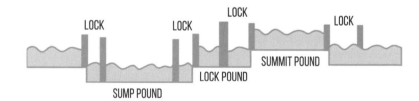

TYPES OF LOCK

There are six types of lock:

Single locks: a single chamber with a **gate at** either end.

Broad locks: single locks that are **broad enough for two boats at a time**, so boats can pass each other.

Double locks: **side-by-side** single locks.

Stop locks: completely **cut off flow** in a canal—used to **prevent rival canal companies stealing each other's water**.

Guillotine locks: **pound locks** with **gates that slide up and down** rather than swing open; used where **space is tight**.

Staircase locks and flights: a series of **locks** enabling the canal **to climb a steep slope**. A **staircase** is where **each gate serves the chamber on either side**; a **flight** is where there is a **pound between each lock**.

ROADS

Roads are probably among the earliest works of civil engineering undertaken by humans, with many roads laid in prehistory before the development of cities or "civilization."

- **c.5000 BC** Ridgeways in use in Neolithic Europe
- **4000 BC** Stone roads in Ur
- **4000 BC** The Corduroy Road (of transverse logs) built in Glastonbury, England
- **4000 BC** At Mohenjo Daro, in the Indus Valley, crushed pottery used to pave streets
- **c.2600 BC** First paved roads in Egypt
- **c.600 BC** Construction of the Diolkos, a stone road across the isthmus of Corinth along which cargoes and even ships were hauled
- **c.500 BC** Construction of the Royal Road in Persia
- **312 BC** Construction of the first stretch of the Via Appia, the greatest Roman road
- **c.AD 50** Claudius completes a road from Britain to Rome
- **c.9th century** The Caliphate develops roads across Islamic lands
- **c.12th century** Medieval Europeans develop a local road network
- **15th century** Development of the Inca road network
- **c.1770** French engineer Pierre-Marie-Jérôme Trésaguet develops a scientific road-laying system
- **1811** John McAdam develops his principles of road laying, "macadamizing"
- **1848** Tar macadam comes into use
- **1858** Asphalt introduced in Paris

ROMAN ROAD CONSTRUCTION

There were at least **three grades of Roman road**: *via terrena* used **packed earth** and *via glareata* used **gravel on packed earth**, sometimes with **flagstones** on top. The **highest grade** was the *via munita*, constructed by **teams of engineers and laborers** (often soldiers), with **deep ditches** filled with **layers of different grades of stones and gravel**, topped with **flagstones cut and laid to give a camber**, and **edged with sidewalks**.

ANCIENT ROMAN ROAD SHOWN IN CROSS SECTION

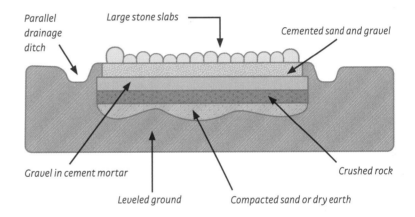

Parallel drainage ditch

Large stone slabs

Cemented sand and gravel

Gravel in cement mortar

Crushed rock

Leveled ground

Compacted sand or dry earth

BETTER ROADS

Improvements in road building pioneered by engineers such as **John McAdam** (1756–1836) **dramatically changed road transport**. The **journey time from London to Edinburgh improved** from ten days in 1754 to just over forty-two hours in 1832.

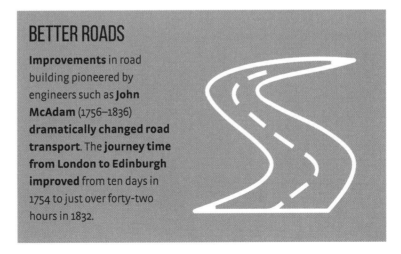

SHIPS

Engineering has always been key to making vessels that can carry loads on water and cope with waves and wind.

c.800,000 BC First maritime voyages by hominids, as *Homo erectus* builds rafts or canoes

c.10,000 BC Twenty-man reed boat shown in prehistoric rock art near the Caspian Sea

c.8000 BC Oldest surviving boat— the *Pesse* dugout canoe

c.4000 BC Large reed boats on the Nile

c.3000 BC Metal tools allow production of planks for larger and more complex vessels

c.2500 BC Egyptians building ocean-going wooden vessels

c.1550 BC The galley is developed by the Phoenicians

c.AD 200 Development of junks in China

c.750 Viking longships

c.1000 Polynesians in outrigger canoes colonize Pacific islands

c.1400 The Portuguese develop caravels

c.1500 Three-masted carracks used for ocean sailing

1769 Early schooners

1774 First paddle steamer

1802 First commercial paddle steamer

c.1830 Early clippers

1843 SS *Great Britain* is the first steamship with an iron hull

1859 First ironclad warship

1886 First oil tanker

1906 HMS *Dreadnought* launches dreadnought era

1918 First aircraft carrier

1951 First purpose-built container ship

1955 First hovercraft

CARVEL VS CLINKER

In ancient times, **two important features distinguished Mediterranean from northern European ships**:

Mediterranean vessels were **carvel built**, with **planks** that **fitted together edge to edge**, to give **smooth sides**; northern European ones were **clinker built**, with overlapping planks, to cope with **heavier seas**.

Mediterranean vessels had **distinct prows and sterns**, whereas northern European ones had **similar profiles at each end**, so that they could **be sailed in either direction as the wind changed**.

CONTAINER SHIPS

Container ships carry **standardized boxes (containers), designed for ease of handling in** loading and unloading and transferring to road and rail carriers. The **more a ship can carry**, the more **economical its transport costs**, especially since even the largest ships **today are effectively robotic** with **minimal crews**. Accordingly, the **size** of container ships **has ballooned**.

OIL TANKERS

An oil tanker is essentially a tank or series of tanks **enclosed in a ship-shaped shell**, with an engine attached to one end. **Superstructures**, such as the bridge, are kept to a minimum, as are crew sizes.

SUBMARINES

A vessel that makes it possible to travel underwater and stay there for long periods had been a dream since the time of Aristotle. Tales of Alexander the Great traveling like this inspired attempts to create submarines.

SUBMERSIBLE VS SUBMARINE

A true submarine is **completely independent of any support from the surface** and **only needs to resurface periodically to refuel**. Most of what are described as historical examples of submarines are actually **submersibles**: vessels that **rely on a connection to the surface** for **air, power**, etc.

PRINCIPLE OF BUOYANCY

The key principle in designing a submarine is **buoyancy**: a body **will sink only if it weighs more than the water it is displacing**. Submarines need to be able to control their buoyancy so that they can **dive** and **surface**. To achieve this, they need some way to **change their weight or volume**.

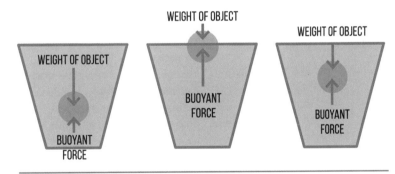

DREBBEL SUBMARINE

Dutch inventor **Cornelis Drebbel** (1572–1633) designed a **contraption often touted as the first submarine**, although in fact it was **barely a submersible**. It was a **covered-over rowing boat** with an **angled prow** that supposedly allowed it to dive as it went forward.

PRINCIPLE OF THE DIVING BELL

A diving bell works because **air trapped under an upturned vessel cannot escape**. Early diving bell users with no **air supply** had to contend with many problems, including **depletion of oxygen** and **buildup of carbon dioxide, heat, and humidity**. Also, **the deeper they went**, the **heavier the bell became**, because, following **Boyle's law**, the **water pressure compressed the air in the bell**, decreasing its **buoyant volume**.

STEAMSHIPS

Until the late eighteenth century, steam engines were huge, lumbering machines limited to mining applications, but the new engines pioneered by James Watt offered a radical new technology: could marine engineers take advantage?

TRANSPORT

1783 The Marquis de Jouffroy d'Abbans builds the first steamboat, the *Pyroscaphe*

1790 John Fitch runs the first steamboat service in America

1801 The Scottish engineer William Symington developed a steam engine to power a small river boat, the *Charlotte Dundas*

1815 USS *Demologos*, the world's first steam-driven warship

1819 The *Savannah*, a sailing ship with a steam engine and paddle wheels, makes the first steam-assisted crossing of the Atlantic

1840 The SS *Great Britain* is the first steamship to use screws

1860 The SS *Great Eastern* becomes the largest passenger ship for the next forty-seven years

1884 Invention of the marine steam turbine

PADDLE WHEELS VS SCREWS

For nearly forty years the **paddle wheel** was the device **powered by steam engines to drive ships through the water**, but **paddle wheels had drawbacks**. In particular they **could submerge** or **heave clear** in **heavy seas**, **damaging the engines**. **Screw propellers**, introduced by **Isambard Kingdom Brunel (1806–59)** on the ***Great Britain*, solved this problem**.

STEAM CANOE

In America, engineer **John Fitch (1743–98)** contrived an **alternative to both wheels and screws**. His first steamboat, the ***Perseverance*, had banks of paddles** on either side **mounted on a crank** so that they **rowed the boat like a canoe**. Later, he switched to **rear-mounted blades** that **paddled like a duck's feet**.

MARINE ENGINEERS

Steam shipping created a **new class of engineer**: the "**marine engineer**," who **replaced the traditional sailor**. Instead of working **sails** and **ropes**, the marine engineer had to **stoke and maintain the engines**.

TRIPLE EXPANSION

The **economics**, **speed**, and scale of **steam shipping** were **transformed** by the introduction in the 1870s of the **triple expansion engine**, in which **steam is used three times to drive pistons before being condensed**. Even **greater efficiency and power** came with the introduction of **steam turbines**, where the **steam drives the rotation of a propeller-like turbine**.

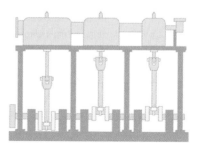

LOCOMOTIVES

Railways existed before locomotive engines, but it was steam engine technology created by pioneering engineers that made them transformative.

1801 Cornish engineer Richard Trevithick's Puffing Devil locomotive pulls six passengers along the main street in Camborne. A few days later, the engine is destroyed when its operators leave the boiler running as they pop into the bar for a drink, and the engine boils dry and overheats

1804 Trevithick's unnamed steam locomotive engine hauls ten tons of iron and seventy men nearly ten miles on the tramway at the Penydarren Ironworks in Merthyr Tydfil, at an average speed of 5 mph. This is the first successful application of a steam engine on a railway

1812 The Middleton Railway near Leeds is regularly hauling freight using a steam train

1813 William Hedley's Puffing Billy begins fifty years of service hauling coal wagons at Wylam Colliery

1814 George Stephenson (1781–1848) designs his first steam locomotive, Blücher, for hauling coal along the tramway at Killingworth Colliery

1825 Opening of the first public passenger and freight railway in the world—the Stockton and Darlington Railway. Stephenson's engine Locomotion No. 1 pulls a carriage named Experiment, preceded by a man on a horse carrying a flag, which reads: *Periculum privatum utilitas publica* ("The private danger is the public good")

WALK, DON'T RUN

On its first trip along the **Stockton and Darlington Railway, George Stephenson's Locomotion No. 1** achieved an **average speed** of 8 mph. The **average speed of a running man** is 10–15 mph.

STEAM CIRCUS

To promote his steam engines, **Trevithick** set up a kind of "**steam circus**," with a **circular demonstration track** in **London's Torrington Square**, around which ran his locomotive **Catch Me Who Can**, which is sketched below.

STEPHENSON'S ROCKET

The most famous of all early steam engines, Stephenson's Rocket opened the door to a new world of speed, thanks to two engineering innovations: multiple boiler pipes and direct drive.

AT THE RAINHILL TRIALS, STEPHENSON'S ROCKET ACHIEVED A TOP SPEED OF 30 MPH.

TRANSPORT

THE RAINHILL TRIALS

George Stephenson's locomotives had **pulled freight** on the **Stockton and Darlington Railway (S&DR)** for several years, but it was felt that a **new design** was **needed** for **passenger service**. In 1829 the S&DR held trials at **Rainhill** to select from competing designs. **George and his son Robert designed the winner**, Rocket.

DIRECT DRIVE

Older locomotives were **beam engines** like those used at **mines and mills**. In a beam engine, the **cylinder is vertical** and the **piston drives a horizontal** beam so that it **rocks up and down**. Stephenson's Rocket had **inclined cylinders** that **drove the wheels directly**. **Direct drive would become standard**, although the **cylinders would eventually be horizontal** for a smoother ride.

BETTER BOILING

Rocket was able to go faster than other locomotives because it could **generate more steam power**, and this was down to its **innovative boiler**. Previously, steam engine boilers had used **one or two wide iron pipes or flues to convey hot gases from the furnace through a body of water to heat it into steam**. Rocket used **twenty-five narrow copper tubes** (two inches instead of twelve), **greatly increasing** the surface area for and rate of **heat transfer** in the boiler.

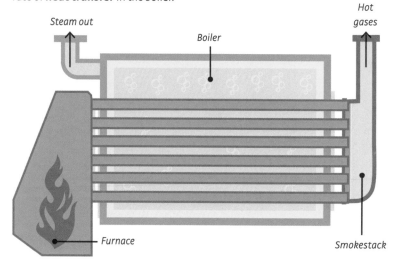

Steam out

Boiler

Hot gases

Furnace

Smokestack

FOUR WHEELS GOOD, SIX WHEELS BETTER

Rocket and all the other steam engines before it had in common a **design flaw**: because they had only **four wheels**, they would **"pitch"** or **seesaw** when moving **at even moderate speeds**. Also, as engines got **heavier**, the **load on the wheels** was **increasing**. To solve these problems, in 1833 the **Stephensons** introduced the **first six-wheeled engine**, the **Patentee**.

OMNIBUSES

Large carriages or coaches were common for private use and for city-to-city travel, but specially designed horse-drawn carriages for public transport in cities were an innovation not introduced until the 1820s.

- **1662** A public carriage transport experiment in Paris fails

- **1816** A short stagecoach service introduced between Manhattan and Brooklyn in New York

- **1826** Stanislas Baudry starts the first horse bus for public transport service in Nantes, coining the name "omnibus"

- **1829** George Shillibeer starts an omnibus service in London

- **1830** An omnibus service introduced in New York

- **1847** First double-decker bus

- **1851** The Great Exhibition in London drives demand for bus services; introduction of knifeboard seating

- **1914** Last horse bus service in London

- **1956** Introduction of London Routemaster bus

ONE FOR ALL

Omnibus means **"for all."** The **generic name for urban road public transport carriages** came from the Frenchman who started the **first successful service**, **Stanislas Baudry** (1777–1830). He called his carriage an "omnibus" because it was a **service for everyone**.

HORSE BUS SEATING

Horse buses had to **accommodate as many people** as possible without diminishing too much the **quality of their ride**. The standard omnibus carriage had front-facing **"garden seating"** for twelve people, with a **ladder that led to the roof**, which served as an uncovered **upper deck** with back-to-back five-seater benches known as **knifeboards**, and **seating for two more next to the driver. Replacing the ladders with stairs** made the **top deck accessible to female passengers**.

OPEN DOOR POLICY

The Routemaster also featured a simple but ingenious strategy to **improve efficiency** and **reduce stoppages**: an **open rear platform**, so that passengers could alight and board anywhere, so long as the bus wasn't moving too fast.

TRANSPORT

ROUTEMASTER INNOVATIONS

The most famous **motor bus** was the **Routemaster**, a **double-decker bus** designed for the **London** market in the years after the Second World War and coming into service in 1956. Engineers had been tasked with producing a bus that was **lighter** (and therefore more **fuel efficient**) than previous models **without sacrificing capacity**. Lightweight **aluminum construction** developed in **wartime aircraft production** led to a **dramatic weight reduction**, and the Routemaster was also the **first bus to feature independent front suspension, power steering**, a **fully automatic gearbox**, and **power-hydraulic braking**.

ISAMBARD KINGDOM BRUNEL

Son of the engineer Marc Brunel, Isambard Kingdom (1806–59) would become one of the most celebrated engineers of all time, with great achievements in bridge, railway, and steamship engineering.

WET WORK

Isambard's **first job**, in 1825, was overseeing the digging of the **Thames Tunnel** with his father's **tunneling shield**, but in 1828 he **nearly drowned** when the **tunnel flooded**.

CLIFTON SUSPENSION BRIDGE

Isambard submitted the winning design for a bridge over the **Clifton Gorge** in **Bristol** and work started in 1831, but the bridge did not open until 1864. It remains an iconic structure to this day.

IRON SHIP

In 1843, Isambard launched the **SS *Great Britain,*** an **iron-hulled steamship** that, as the *Great Western* had been, was the **largest steamship yet built**. It was the **first to use a screw propeller instead of a paddle** and was in service for decades.

GO WEST

In 1833 Isambard was asked to build the **new railway line between London and Bristol** (and beyond), the **Great Western**. He surveyed the route, oversaw the laying of 118 miles of track, and designed **rails**, **tunnels**, **bridges**, **stations**, **signals**, and even **station lampposts**.

GO WESTER

In 1835, Isambard suggested that the **Great Western line** (as opposed to the physical train track) should not stop at the sea but **continue across the Atlantic**, and he was **commissioned to build the first steamship to make the crossing under steam alone**. At the time, it **was claimed** that to do so **would require the ship to be filled with coal**, but the **SS *Great Western*** proved this wrong in 1838.

THE GREATEST

Isambard's **final project** was a huge steamer nearly ten times heavier that the *Great Britain*. The ***Great Eastern*** was designed to carry **enough coal to get to Australia**, and its visionary **double-skinned iron hull** would become **standard for steamships**. But it was a **troubled project** and Isambard had a **fatal stroke just two days before the ship's maiden voyage** in 1858.

BICYCLES

A two-wheeled vehicle propelled by its rider, the bicycle is such a pinnacle of engineering that the basic design has changed little since the 1880s.

c.1790 First hobbyhorse-style two-wheel vehicles

1817 "Dandy-horse" or "swift-walker" steerable bicycle

1839 First pedal bicycle

1861 "Boneshaker" velocipede with pedals on front wheel

1871 Penny farthing

1873 Invention of chain drive

1885 Modern safety bicycle

PAVING THE WAY

Bicycle engineering and manufacture paved the way for the rise of the automobile. Bike manufacturers pioneered techniques such as the **assembly line**, **planned obsolescence**, and **marketing incentives**. **Cycling organizations** agitated for **good-quality roads** and **effective regulations** (speed limits, traffic controls, etc.), **which facilitated increased car use**.

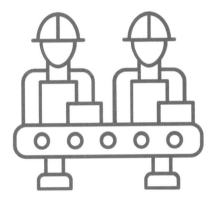

THE GOLDEN AGE

The **safety bike**, with **chain drive, brakes, pneumatic tires**, and ball bearings, made the **bicycle faster than a horse and carriage**, but easy and safe to ride. In the United States it led to a **"golden age of cycling."** In 1887 there were a **hundred thousand bicycles on the road** in the United States; by 1896, there were over **four million**.

CYCLING MALADIES

The craze for cycling prompted warnings of the **moral dangers of the bicycle**—which was held to encourage in female riders, for instance, risqué clothing such as bloomers—and **confected health risks** such as **"bicycle hump"** and **"bicycle twitch."**

ELEVATORS

The safety elevator was an ingenious life-saving invention in its own right, but it also had profound effects on architecture, urban life, and property values.

TRANSPORT

GOING DOWN

Passenger elevators were **introduced in Britain** in the 1830s **and the United States** in the 1840s, but the **hemp ropes** used **often broke**, with **fatal results**.

OTIS TO THE RESCUE

In 1852, engineer **Elisha Otis** (1811–61) was in charge of refitting a factory and noticed that the workmen steered clear of the **hoist elevator**. He resolved to improve its safety and so his **safety elevator invention** was born.

UPSIDE DOWN

The advent of elevators abruptly **reversed the normal pattern of property values in high-rise buildings**. Before the elevator, the **rooms at the top** of the building were the **cheapest** because **hardest to reach**. **Afterward, the top floor**, away from **street smells and noise, became the most expensive**.

PAWL BEARERS

Otis's basic innovation was to **run saw-toothed ratchet-bar beams** down each side of the **elevator shaft**, with **pawls** on the elevator that **automatically engaged with the ratchet unless under tension**. **If tension failed**, as when the **elevator cable broke**, the **pawls would spring out** to the side and **stop the elevator**.

HIGH PLACES

Although Elisha Otis died in 1861 aged just forty-nine, his **sons built up his business**, installing **Otis elevators** in Paris's **Eiffel Tower** in 1889, the **Washington Monument** in 1890, and inside the **Woolworth Building**, the **world's tallest** at the time, in 1913.

ELECTRIC RAILWAYS

Werner Siemens's dynamo-electric power generation made possible new applications of electric power, including transport.

NEED FOR CLEAN

In the mid-nineteenth century, **growing demand for urban transport** could not be adequately met by the existing technologies: **horse and steam power**. The **former** was **not powerful enough**, while the **latter** was **too dirty and cumbersome** for **urban contexts**. **Electric motors** might offer the **clean, scalable motive force required**, if enough power could be made available.

ELECTRIC EXPERIMENTS

Experiments with electric motors for transport dated back at least as far as 1835, when US blacksmith **Thomas Davenport** demonstrated a **small electric train**. But **battery power** was then **insufficient**, and although **delivery of current by rail** was patented in 1840, there was still **no sufficient source of power available**.

ELECTRIC EMPIRE

Siemens followed this success with the **first electric tramway in Berlin**, in 1881, and went on to devise the **first electric trolleybus, electric mine locomotives**, and **electric underground railway** (in Budapest).

JOY RIDE

This changed in 1866 when **Werner Siemens** invented the **self-exciting dynamo**. Seeking new applications for this, Siemens devised an **electric train demonstrator** for the **1879 Berlin Industrial Exposition**. A small **electric-motor-powered train** pulled bench-style seating for **eighteen passengers** around a 1,000-ft.-long circular track. **Electricity** from a **steam-powered dynamo** was **fed to the train** through the **rails** of the track. The little train carried more than 86,000 passengers in four months.

INTERNAL COMBUSTION ENGINE

Rapidly to become the most popular engine in the world when converted to work with petroleum, the original internal combustion engine ran on gas.

IN OR OUT?

An internal combustion engine (ICE) is so named because **fuel burns inside the engine itself**. A steam engine, on the other hand, **is an external combustion engine**, because the **fuel is burned in a furnace separate from the engine**.

OTTO'S ENGINE

German engineer **Nikolaus Otto** (1832–91) created the **first working ICE** in 1876. It was a **four-stroke engine** that ran on **gas**.

FOUR-STROKE CYCLE

The **cylinders** in Otto's engine ran on a four-stage cycle, where **each stage was accompanied by one motion of the piston**. **Valves** were used to **control the flow of gases** in and out of the cylinder at each stage.

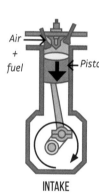
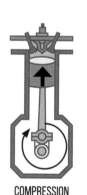
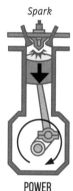
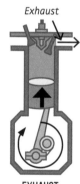

Air + fuel — Piston

Spark

Exhaust

INTAKE   COMPRESSION   POWER   EXHAUST

TWO-STROKE CYCLE

In a **two-stroke engine**, the **piston itself acts as the valve** to **shut off intake or outlet ports**, which reduces complexity.

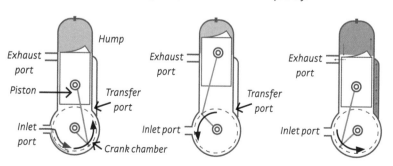

Hump

Exhaust port

Piston

Transfer port

Inlet port

Crank chamber

Exhaust port

Transfer port

Inlet port

Exhaust port

Inlet port

OIL BOOM

The invention of the **carburetor** at the end of the nineteenth century **meant the ICE could run on petroleum**, just in time to coincide with the rapidly developing **oil industry**.

AUTOMOBILES

A self-propelling road vehicle, or horseless carriage, the automobile or motor car could become a commercial reality only when engine technology reached a certain level.

c.1770 First working automobile—steam powered—built by French inventor Nicolas-Joseph Cugnot

1807 First vehicle with an internal combustion engine built by Swiss inventor François Isaac de Rivaz

1832 First electric vehicle

1876 Nikolaus Otto's steam-driven internal combustion engine

1879 German engineer Karl Benz receives patent for petroleum ICE

1885 Benz's *Patent-Motorwagen*—a three-wheeled vehicle—becomes the first successful automobile

1886 Gottlieb Daimler builds the first four-wheeled car

1896 Rudolf Diesel designs the diesel engine

BODY PLAN

The **basic body plan of the motor car** is **unchanged since it was invented**. There is a **chassis**, to which are connected the **engine** or power plant; the **transmission system** to transfer drive to the **wheels**; **and suspension**, **steering**, **and braking**.

BENZ PATENT MOTOR CAR

In 1879 Benz had built a **stationary one-cylinder two-stroke ICE**, the commercial success of which financed his development of the **Patent Motor Car (*Patent-Motorwagen* in German)**. The car had a compact **high-speed single-cylinder four-stroke engine installed horizontally at the rear**, together with several features borrowed from contemporary bicycle engineering, including a **tubular steel frame** and **wire-spoked wheels**. The **engine output** was just 0.75 hp. (0.55 kW).

FORD AND THE ASSEMBLY LINE

Henry Ford (1863–1947) was a successful designer of motor cars who started his own company in 1903 and came out with his Model T in 1908. But it was his 1913 assembly line that revolutionized the automobile industry—and most others.

MODEL T

Ford used all his experience to create **a car for everyone**, culminating in his 1908 Model T. With its **tough chassis**, **high wheel clearance**, and **simple design**, it could **cope with poor roads** and was **easy for owners to fix**. But it was **expensive**.

HIGHLAND PARK

Ford constructed a massive plant at **Highland Park** in **Michigan**, and combined several innovations. From **slaughterhouses** he borrowed **assembly lines**, with objects suspended from **chains** moving between **work stations**; from **grain warehouses** and **mills** he borrowed **conveyor belts** and **gravity slides**.

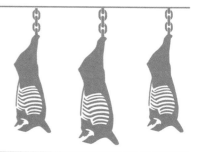

STEPS AND PARTS

Following the principles of **Taylor's scientific management**, Ford broke down the manufacture of the Model T into **eighty-four steps**, and ensured that **all the parts** were **standardized** and **interchangeable**. The chassis moved along the line between **worker stations**, where **specialization** improved **efficiency** and **quality**.

FASTER AND CHEAPER

The **time it took to produce a single Model T** dropped from twelve hours to ninety-three minutes. In 1914 Ford produced 308,162 cars—**more than all the other automobile manufacturers combined**. The price of the Model T **dropped** from $850 in 1908, to $316 in 1916, to $260 in 1924 (in modern terms, from $21,000 to $3,500).

HOVERCRAFT

*In the 1950s, British engineer Christopher Cockerill created
a new class of vehicle that uses a cushion of air to reduce friction.*

SMOOTHER SAILING

Cockerill helped develop **radar** during the Second World War before retiring to run a marina. There he mused on ways to make **water travel more efficient** by reducing friction, the force that opposes motion for any vehicle moving across a surface.

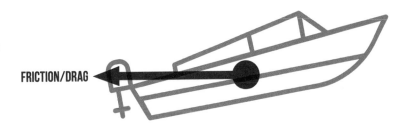

FRICTION/DRAG

TIN-POT THEORY

Cockerill developed the idea, first explored by Victorian engineer **Sir John Thornycroft** in the 1870s, of reducing friction using a cushion of air. He experimented using a **cat food tin inside a coffee can**; when he blew air through the cans, the inner one floated.

HOW IT WORKS

A hovercraft uses a **powerful fan** to blow air beneath the vessel, which usually has a **skirt of rubber or fabric** to contain the **pressurized air**. This air cushion reduces friction and **lifts the vessel clear of waves or an uneven land surface**, making it much easier to drive it forward (with the use of a lateral air stream).

BY LAND AND SEA

Hovercraft are particularly suited to **amphibious applications**, since as long as the surface is relatively level, they happily **travel over land, marsh, mud**, etc., **as well as over water**. Thus they are especially used by the **military** and those traveling in **mixed land and water environments** such as the **Florida Everglades**.

CHANNEL CROSSING

With the help of a British government agency, Cockerill developed the **first hovercraft**, launched in 1959. It **crossed the Channel** between England and France a few weeks later.

SELF-DRIVING CARS

Engineering the car of the future involves the integration of a suite of technologies.

CHALLENGES

The four key challenges facing a self-driving car are:

Environmental perception—knowing what's around.

Path planning—working out a route through the environment.

Car control—controlling the functions of the car.

Navigation—plotting a route from A to B.

SENSORS

A range of **sensors** are available to help the vehicle achieve **environment perception**, including **infrared** and **ultrasound**, **radar** and **lidar (laser ranging)**. Some systems use **visual recognition** via **cameras**. An emerging technology is **V2X** or **vehicle-to-everything communication**, where the vehicle **gets information from other cars**, **smart street furniture** (such as **road signs**), etc., via **wireless communication**.

SMART CARS

To crunch all the data about **environmental perception**, **location**, and **navigation**, and use this to make **car control decisions**, vehicles need **data processing**, both **on-board** and **cloud-based**, with sophisticated **deep-learning artificial intelligence**.

CAR IQ

Self-driving is classified into levels:

Level	Judgment standard
No-automation (Level 0)	The **driver completely controls the vehicle** all the time.
Function-specific automation (Level 1)	**Individual vehicle controls** are **automated**, such as **electronic stability control** or **automatic braking**.
Combined function automation (Level 2)	At least **two controls** can be **automated in unison**, such as **adaptive cruise control** in combination with **lane keeping**.
Limited self-driving automation (Level 3)	The **driver can fully cede control of all safety-critical functions in certain conditions**. The **car senses when conditions require the driver to retake control** and provides a "**sufficiently comfortable transition time**" for the driver to do so.
Full self-driving automation (Level 4)	The **vehicle performs all safety-critical functions for the entire trip**, with the **driver not expected to control the vehicle at any time**. As this **vehicle would control all functions** from start to stop, including all parking functions, it **could include unoccupied cars**.

SPECTACLES AND LENSES

Visual impairment must have dogged humans since they first evolved, but only with the application of optical engineering in the Middle Ages did it become possible to correct this.

- **c.AD 1000** Arab scholar and astronomer Ibn al-Heitam suggests that shaped lenses might correct visual impairment

- **c.AD 1000** A lens maker on the Baltic island of Gotland grinds high-powered lenses, possibly for use as "reading stones"

- **1266** English monk Roger Bacon sets out the scientific principles of corrective lenses

- **c.1280** Venetian glass makers put two lenses together in a frame to make the first spectacles

- **c.1720** First temple (ear-loop) glasses made in London

- **1784** Benjamin Franklin creates bifocal lenses

REFRACTION

Lenses work by **refracting light rays**, which means making them **change course**. Light refracts when it passes **from one medium to another**—for instance, from air to glass—because, as it does so, **it changes speed**. A good analogy for understanding this is with wheels on an axle passing from smooth ground to rough: the first wheel to hit the rough will slow down and so the whole axle will turn slightly in that direction.

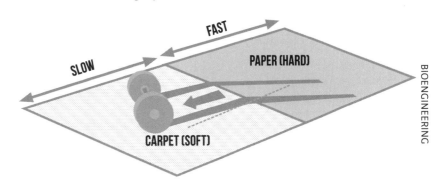

FAST

SLOW

PAPER (HARD)

CARPET (SOFT)

NEAR AND FAR

The human eye has a **natural lens** that is supposed to **focus light on the retina**, but faults can develop that shift the focal point in front of (**near-sightedness**) or behind (**far-sightedness**) the retina. Lenses correct such impairment by **changing the focal point of light rays**.

NEARSIGHTED

CORRECTED NEARSIGHTED

THE EYE

NORMAL

FARSIGHTED

CORRECTED FARSIGHTED

LIGHT

ECG

The electrocardiogram (ECG) was a major advance in medical technology, and a landmark in the application of electrical engineering to medicine.

ANATOMIES

Physicians in the nineteenth century were making great advances in understanding the workings of the body, especially by **cutting open the bodies of live animals and dead humans**.

A WINDOW ON THE HEART

What they needed more than anything, however, was a technology that could let them know what was going on inside the bodies of **living patients**. Listening devices (such as the **monoscope**) offered **a way to listen in on the heart**, but doctors needed more.

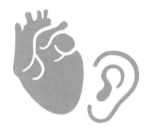

SCRATCHING AROUND

An instrument devised in 1860 made a record of the **heart's pulses**, using **sound vibrations to scratch marks on smoked paper**. In 1887 **Augustus Waller** became the **first to record the electrical activity of the heart**, but the device he used was inaccurate and primitive.

HEART STRINGS

In 1903 Dutch physician **Willem Einthoven** invented the **string galvanometer**, a very sensitive device for **measuring electrical activity**, consisting of a metal wire strung between two electromagnets. Wired up to the arms and legs of a person, it **could pick up the electrical signals generated by the heart**, as they spread through the person's skin.

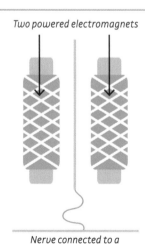

Two powered electromagnets

Nerve connected to a gold-coated string

ANATOMY OF A HEARTBEAT

Einthoven managed to get his machine to produce a **standardized, consistent, and recordable trace on paper: an electrocardiograph**. He was even able to break it down to its main components, labelled, P, Q, R, S, T, and U. P is the electrical wave associated with **atrial contraction**, while the others are associated with **ventricular contraction**.

IRON LUNG

Polio could lead to terrifying death by suffocation, but engineering came to the rescue, with an initially crude device that made use of basic scientific principles.

POLIO PARALYSIS

Now all but eliminated in the Western world, polio is a disease that can cause **paralysis**, including of the chest muscles, which **restricts breathing** and can lead to suffocation. In the 1920s, hospitals were desperately seeking a way to keep patients alive long enough for life-saving treatment.

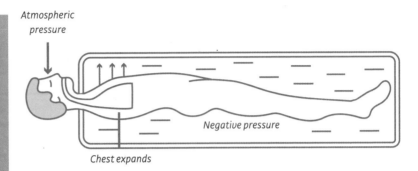

Atmospheric pressure

Negative pressure

Chest expands

ENPV

The respirator worked through a mechanism called **External Negative Pressure Ventilation** (ENPV). The principle is that if the body of a patient, below the neck, is sealed in a box in which the pressure is reduced (by sucking air out with bellows), the **atmospheric air pressure** outside the box will **force air to flow into the lungs**. Blowing air back into the box causes the pressure inside to rise, so that the patient's **lungs deflate passively**, causing them to **breathe out**.

CURARE CAT

In 1927 **Philip Drinker** and **Louis Agassiz Shaw**, both of Harvard University, injected a cat with the paralyzing poison curare and showed that they could keep it alive with an **artificial breathing mechanism**, or **respirator**, involving a sealed box and a set of **bellows. This was the basis for the first iron lung design**.

BIG BOX

The first ENPV device, a.k.a. the **iron lung**, was a very large metal box with a mechanical bellows attached. Other engineers later modified the design to make it easier to access the patient, and even made **lightweight, cheap, and easily manufactured plywood versions**.

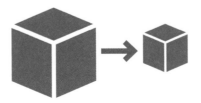

ROCKING BEDS

Rehabilitation helped patients to **strengthen their chest muscles and start breathing on their own again**. One simple trick to help them was a **rocking bed**, which **tilted the patient** up so that their internal organs shifted down, helping to **draw air into the lungs**, and then tilted the other way to reverse the effect.

ARTIFICIAL HEART VALVES

Prosthetics are at least as old as civilization, but to replace the hardest-working component of the hardest-working organ in the human body would be a bioengineering feat of breathtaking audacity.

HEART VALVES

The heart is an organic pump with **four chambers**, the job of which is to **draw in oxygenated blood** from the lungs, **pump it out** to the rest of the body, **receive it back**, and **send it to the lungs** to begin the cycle anew. The heart relies on valves between its four chambers to ensure that blood **flows in the right directions only**.

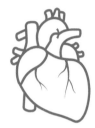

VALVE DEFECTS

Faults in the valves can cause them to become **leaky**, threatening both the **integrity of the heart** and the **blood supply** to the rest of the body. Heart surgeons dreamed of being able to replace such valves.

BALL AND CAGE

Following a series of **animal trials**, in 1952 **Dr Charles Hufnagel**, Professor of Experimental Surgery at Georgetown Medical Center in Washington, D.C., implanted a **caged ball valve** in a patient with **aortic valve disease**.

TILTING DISCS

Following this surgery, there were **multiple refinements of the ball valve design**, while in the late 1970s **tilting disc and leaflet designs** were introduced. These are now made of pyrolytic carbon (originally used in the nuclear industry).

BALL AND CAGE

TILTING DISC

BILEAFLET

LASTING POWER

The great advantage of **mechanical heart valves** is their remarkable **durability**. Some of the ball valves lasted for over thirty years without significant wear, while over 2 million **tilting disc valves** have been deployed and there have been **virtually no reports of mechanical failures**, ever. The downside is that implant recipients have to take **blood-thinning drugs** for the rest of their lives to ensure that **blood clots** don't block up the valves.

HEART-LUNG MACHINE

Complex life-saving surgery on the heart is only possible if there is some way for the body to cope without it during the operation; for these vital minutes the heart-lung machine keeps the patient alive.

MULTI-PURPOSE

The heart-lung machine (HLM), also known as the **cardiopulmonary bypass pump** or **extracorporeal perfusion circuit**, has more than one job to do. It **must keep blood circulating around the body**, standing in for the heart, and it must also stand in for the lungs by **oxygenating the blood and removing carbon dioxide**, while **maintaining the temperature of the blood** and **ensuring it does not clot**.

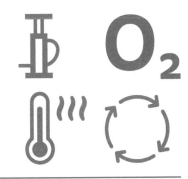

ANIMAL RESEARCH

In 1931 surgeon **John Gibbon** was dismayed when a patient died under the knife because their **circulation failed** while their heart was being operated on. He and his **wife Mary** launched a twenty-year research program to make a machine that could **artificially maintain circulation**, testing it on **cats and dogs**. In 1953 Gibbon used the machine for the first time, successfully operating to repair a heart defect.

ROCKY ROAD

Although the machine represented a massive breakthrough, it had many problems; its processes **damaged blood cells** and introduced **dangerous contamination to the blood**. Not until the development of new **membrane technology**, to **protect the blood while oxygenating it**, was the machine truly finished.

HOOKED UP

In cardiac surgery a patient is hooked up to the HLM, which is actually a **battery of pumps and other devices**, including **oxygenators** and **filters** to make sure no blood clots get through to the patient.

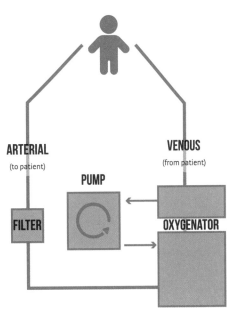

ARTERIAL
(to patient)

VENOUS
(from patient)

PUMP

FILTER

OXYGENATOR

PACEMAKER

When evolutionary electrical engineering develops faults, human electrical engineering can step in and correct them. Nowhere is this more clearly demonstrated than with the pacemaker.

FAULTY WIRING

Correct operation of the heart requires **synchronization of its muscles** so that they **squeeze the right chambers in the right order**. The heart has a built-in pacemaker, called the **sinoatrial node**, to orchestrate this process, but when it goes wrong, dangerous **arrhythmias** can result.

Sinoatrial node

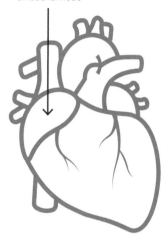

SHOCK TREATMENT

Shocks applied externally can reset the **rhythm of the heart** and even **restart it** if it stops; this is the principle behind **defibrillation machines** or banging someone on the chest. **Electrical shocks** applied directly to the heart can be even more effective, which is why in 1952 the **first cardiac pacemaker** was developed, but it was a **large machine** that had to be carried around on a trolley.

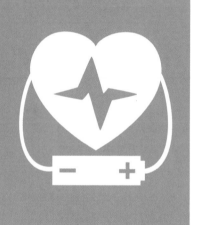

PACEMAKER BREAKDOWN

A modern pacemaker is about the size of a large coin, and is comprised of a **battery**, a **tiny computer**, a **generator**, and **wires with sensors**. The sensors monitor the activity of the heart and feed information back to the computer, which decides when to **generate a stimulus** that passes back down the wires to the heart muscle.

CARDIAC EMERGENCY

In 1958 **Else-Marie Larsson** was desperately searching for a treatment for her husband, who had a **heart condition that made his heart stop** up to thirty times a day. She contacted Swedish inventor **Rune Elmqvist** and got him to provide the world's **first implantable pacemaker**, which was wired directly to the heart to restart it when necessary.

LEADLESS PACEMAKERS

The **latest generation** of pacemakers are **tiny cylinders** that do not require leads and sit **directly in the heart**.

BONE REPAIR

*Bones are the body's structural engineering; to repair them when they fail,
who better to turn to than engineers?*

RODS, PLATES, SCREWS, AND PINS

Elements used to help bones to repair themselves, or to stand in for the structural elements of bones, include:

- **rods**: to **add strength** to long, large bones such as the femur.
- **plates**: to **hold together fragments** of bones.
- **screws**: to **fasten plates** and rods in place.
- **pins**: to **affix detached pieces of bone**.

INTERNAL OR EXTERNAL

Biomedical engineers and physicians work together to determine whether it is better to provide **internal fixation**, by implanting elements like rods and screws, or **external fixation**, which encompasses external supports that are screwed into the bone, as well as **slings**, **splints**, **casts**, etc.

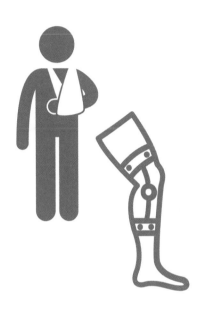

IMPLANT RISKS

Engineering expertise can also warn of **potential dangers from metal implants**. For instance, plates or screws that carry too much load can cause **bone atrophy** through **stress shielding**. Using **dissimilar metals** that end up in contact with one another can cause **galvanic corrosion** and **mobilize toxic metal ions**.

IMPLANT MATERIALS

Materials engineers advise on the best materials to use for **orthopedic implants**. For instance, different **combinations of nickel**, **chrome**, and **molybdenum** produce **stainless steels** with **differing properties**—e.g., more or less flexibility.

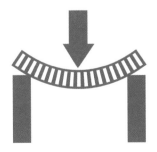

BIODEGRADABLE

Often an implant is intended to be **temporary**, allowing bone to **regrow**, but removing an implant brings its own **risks**—for instance, of **infection**. **Biodegradable polymers** might soon be engineered to **degrade at just the right rate**, **transferring load** to the healing bone over time.

ARTIFICIAL JOINTS

Artificial joints have to be light, hard wearing, durable, and mobile, while also overcoming challenges such as attachment and biocompatibility.

BIOENGINEERING

HEAVY DUTY

Joints have to bear **great loads** while also providing **mobility**. They are enclosed within **capsules** that lock in **lubricating fluid**, while the surfaces of joints are coated with **cartilage** for **lubrication**, **cushioning**, and **bone protection**. When this cartilage **wears away** and/or the joint bones themselves start to deteriorate, **joint mobility** is impeded, and it can be very painful.

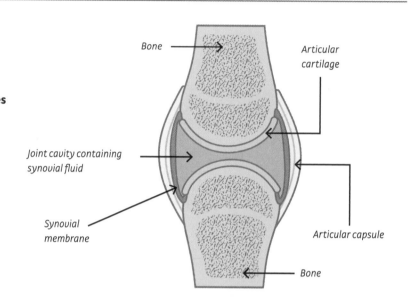

Bone

Articular cartilage

Joint cavity containing synovial fluid

Synovial membrane

Articular capsule

Bone

IVORY AND STEEL

The **first attempt to implant an artificial joint** was made by German physician **Themistocles Gluck**, in 1890. He made an **artificial knee joint** from ivory and **nickel-plated steel**, soon followed by an **ivory hip joint** screwed into place with **nickel screws**. Although these were a short-term success, patients' bodies soon **rejected** the implant materials because they were **not biocompatible**.

JOINT CHIEF

The **first modern artificial joint** was implanted in 1958 by British surgeon **John Charnley**. He made a joint with a **round metal head** that fitted into a **socket made of Teflon**, although this was **later replaced with polyethylene**. Charnley used an **acrylic cement**, **polymethylmethacrylate**, which is still standard today, although now it usually has added **antibiotics**.

TYPES OF JOINT

Different parts of the body require different **degrees of movement**; hence the body has **many different types of joint** that may need replacing.

BIONICS

Biological electronics, known as bionics, are common in science fiction as a means to augment human capabilities. In real life, engineers are still battling to develop bionics that can come close to matching nature's engineering.

CYBERNETICS

"Bionic" is a contraction of "biological electronics," and is related to the word **"cyborg,"** which is short for **"cybernetic organisms."** Both words come from the science of **cybernetics**, which is the **study of control and communication in living things and machines**.

FEEDBACK

Cybernetics looks at how **feedback from the environment interacts with systems to guide and control them**. A classic example is how the structure and position of **legs** influence the **way that animals walk**.

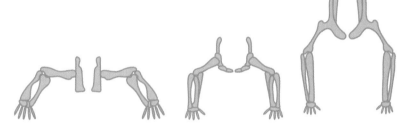

BIONIC PROSTHESES

The dream of bionic prostheses is for **electronics** and **machine parts** to **replace bones**, **muscles**, and **nerves**, to create, for instance, **robotic hands** that can **grasp** and **feel**. The challenges that need to be overcome include **miniaturization of components**, **power supply**, adequate **computer processing**, **viable attachment** of the prosthesis to the body, and **communication between the user and the prosthesis**.

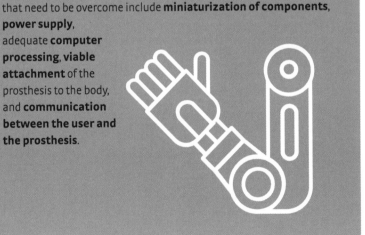

THOUGHT CONTROL

The ultimate goal is for the user to be able to **control bionic body parts in the same way as real ones**: via **nerve impulses** to and from the brain.

COCHLEAR AND RETINAL IMPLANTS

Bionic engineering can help create prosthetics for sense organs. Cochlear and retinal implants are two successful examples of prosthetic aids for the ear and eye, respectively.

BIOENGINEERING

AURAL BYPASS

Most **hearing aids** work by **amplifying sounds** to make it easier for the ear's internal sound-processing structures to pick them up. A **cochlear implant bypasses some of these structures** altogether.

BROADCASTING HOUSE

A cochlear implant has external and internal parts. The external ones include a **microphone** that **picks up sound**, a **processor** that **converts the sound into an electrical signal**, and a **radio transmitter** that **transmits the signal to a receiver** implanted beneath the skin.

DIRECT STIMULATION

The receiver picks up the radio broadcast, converts it into electrical signals, and passes these to **electrodes implanted in the cochlear**, where they **directly stimulate the auditory nerves**.

EYE SEE

Retinal implants involve direct stimulation of the **optic nerve. Light-sensing chips**, either in glasses or implanted in the eye, **convert light into electrical impulses**, which are fed to the optic nerve. Implant recipients can **train themselves** to **interpret this input** as a form of **crude black-and-white vision**.

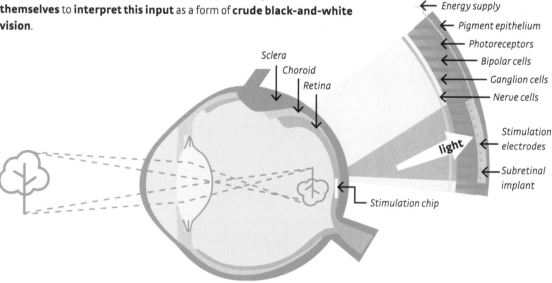

GENETIC ENGINEERING

Also known as genetic modification or manipulation, genetic engineering means tinkering with the genome—the basic biological blueprint—of an organism using biotechnology.

DNA

Short for **deoxyribonucleic acid**, this is the molecule that carries the **genetic code**. Humans have been changing the DNA of animals for as long as they have practiced **selective breeding**, but genetic engineering involves **direct modification** of the genome by **adding or removing pieces of DNA**.

THE CODE

The world of genetic engineering was opened up in 1953 when **Francis Crick** and **James Watson** worked out the **structure of DNA** and showed how the **sequence of elements** called **bases** could **encode genetic information**. Biologists realized that if they could **rewrite the code**, it would be possible to **redraw the blueprint of a living thing**, creating a genetically modified organism.

CUT AND PASTE

The basic steps of genetic engineering are as follows:

create new DNA sequence, ready to paste into host genome

↓

locate gene in host DNA

↓

cut DNA

↓

paste in place the new sequence

↓

reattach cut ends of DNA

↓

host organism reads new DNA as if part of its own genome.

TOOLS FOR THE JOB

In order to start engineering the genome, biologists needed the **right tools** for **reading**, **making**, **cutting**, and **pasting** DNA. Many of these are **borrowed from nature**, by using **naturally occurring enzymes** (nanoscale biological machines). For instance, **restriction endonucleases** cut DNA at specific points, allowing **new DNA fragments** to be inserted.

GENES AWAY!

A major engineering challenge is delivering new genes to the right place. Human DNA, for instance, is hidden away **inside a nucleus inside the cell**, behind **multiple layers of defense**. Methods of **gene delivery** include **blasting genes** into **target nuclei** on **tiny metal particles**, or **co-opting viruses**, many of which have evolved to **inject their own genomes into those of a host**.

CRISPR-CAS9

The biggest development in genetic engineering in decades is a new gene-editing toolkit called CRISPR-Cas9, which makes it easier than ever to snip DNA and insert new pieces.

BACTERIAL TOOLKIT

CRISPR stands for "**Clustered Regularly Interspaced Short Palindromic Repeats**" and Cas stands for "**CRISPR-associated enzyme.**" CRISPR is a **technique evolved by some bacteria for protection against viruses**.

REPURPOSED

Genetic engineers use the CRISPR-Cas9 system for their own ends. They create **short stretches of RNA** (a molecule very similar to DNA, which uses the same code) that **match specific regions of the target organism's DNA**. Scissor-like Cas9 enzymes pick up these RNA pieces and use them to **snip the target DNA at exactly the right spot**.

NATURAL REPAIR

CRISPR-Cas9 **does only half the job**, however. Once the target genome has been cut open, gene engineers rely on the **cell's own DNA repair mechanisms** to stitch the cut ends together again. This is the point at which **modifications** are made, either by **homologous directed repair** (HDR) or **knock-out mutations**.

HDR

Gene engineers make a stretch of DNA comprising a **new "insert" region**, flanked on either side by **homologous DNA sequences** that match the cut ends of the target DNA. These homologous sequences **direct the repair machinery to incorporate the insert into the target DNA**.

KNOCK-OUTS

When a gene is repaired after being cut with Cas9, **errors are often introduced— even one** such error is **enough to knock out or disable the target gene.**

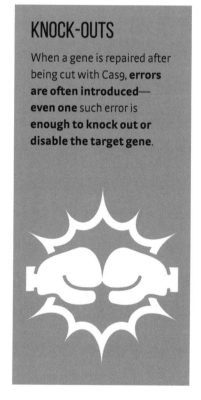

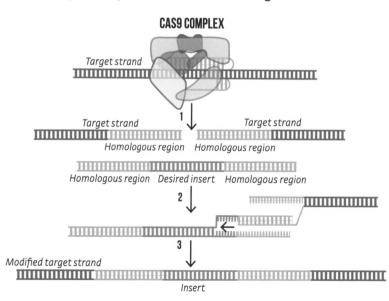

CAS9 COMPLEX

Target strand

Target strand

Target strand

Homologous region   Homologous region

Homologous region   Desired insert   Homologous region

Modified target strand

Insert

BIOENGINEERING

MEDICAL IMAGING

In the nineteenth century, physicians dreamed of opening a window into the living body, so that they could make or confirm diagnoses without having to wait until the patient was dead. Today medical engineering has opened many such windows.

X-RAYS

High-energy, **short-wave electromagnetic radiation**, X-rays can pass through many materials, including **soft tissue**, and this can be used to take an X-ray photograph (a.k.a. **radiograph**) of **bones** and some other tissues. X-rays are not helpful for imaging soft tissue, and their main drawback is that they cause **ionizing radiation damage**, so that **exposure must be limited**.

ULTRASOUND

The **safest type of medical imaging**, ultrasound is a kind of **echolocation**, sending **high-frequency sound waves** into the body and picking up the echoes to generate an image. **Doppler ultrasound** measures the change in the **pitch of sound waves** when reflecting off moving bodies such as circulating blood cells, and can be used to **image blood flow**.

CT OR CAT SCANS

Computed tomography (CT) or **computed axial tomography** (CAT) combines **multiple X-rays** taken from different angles around a single plane to build up a **detailed cross section** or slice through a patient's body. Since it uses X-rays, it involves exposure to **ionizing radiation**.

MRI SCANS

Magnetic resonance imaging uses powerful **magnetic fields** to make the nuclei of **hydrogen atoms** in the body (present in water) act like **tiny radio transmitters**, and the signals from these can be used to generate **very detailed pictures of all types of tissue**. MRI scanners are large, very expensive, and extremely noisy, and cannot be used on patients with **metal implants**, **replacement joints**, or **pacemakers**.

PET SCANS

Positron emission tomography scans for radiation from a **radioactive tracer dye** introduced into the patient. Different dyes can trace different functions; for example, a **glucose-based dye** will show up in **cells and tissue that are using more energy**. PET scans are **often combined with CT scans**. PET scans are **expensive** and involve exposure to **ionizing radiation**.

TISSUE ENGINEERING

Tissue engineering is the science of developing biological substitutes that can maintain, restore, or improve the function of tissues in the body.

REGENERATIVE MEDICINE

Tissue engineering is a vital element in the field known as **regenerative medicine**, and the two terms are often used interchangeably. Regenerative medicine aims to treat patients by helping them to **self-heal**, sometimes with the aid of **engineered tissues**.

TISSUES

The human body is made up of **cells**, which are organized into groups or collections known as tissues (e.g., **muscle tissue**, **cartilage**, **adipose tissue**). Organs are functional units composed of **multiple tissue types working together**.

SUPPORT SERVICES

Cells usually make their own **support structures**, secreting a **network of biochemicals and structures** that make up an **extra-cellular matrix**. In tissue engineering this matrix is known as a **scaffold**, and it serves many functions, including **supporting and nurturing cells**, **guiding their development and growth**, and **passing on biochemical signals**.

MAKING SCAFFOLDS

Tissue engineers seek to make **substitute tissues** by providing **scaffolds**, either **artificial** ones (e.g., made of **plastic**) or ones derived from **natural sources** (such as **cellulose** or **collagen**). With the correct **growth medium** and **cocktail of biochemical signals**, the scaffold can guide the growth of an **engineered tissue**.

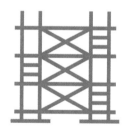

REPLACEMENT PARTS

The ultimate goal of tissue engineering is to grow **new organs from a patient's own cells**, which can be used to **replace failing ones such as kidneys or livers**. Clinical applications so far include relatively minor tissues, such as **small arteries**, **skin grafts**, and **cartilage**.

TISSUE SENSORS

Another goal for tissue engineering is to grow tissues that can be used as **biological sensors**, or integrated—e.g., used as a component in a lab on a chip. Such sensors or **lab agents** can improve **drug testing** or be used in **personalized medicine**.

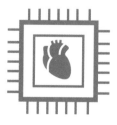

ARTIFICIAL LIFE

Artificial life, often called AL in comparison with AI, is the quest to engineer an entirely synthetic life form.

BUILD YOUR OWN

Ever since the DNA code was cracked and engineers began to make and stitch together their own DNA in the lab, researchers have speculated about **writing their own genome**.

BASES AND CHROMOSOMES

DNA is made up of sequences of units called **bases**; these bases correspond to the **letters of the DNA code**. In organisms, **long stretches of DNA**, encoding hundreds, thousands, or millions of **genes**, are **bundled up with proteins to create chromosomes**. Artificial life engineers seek to compile their own sequences of bases to create **synthetic chromosomes**.

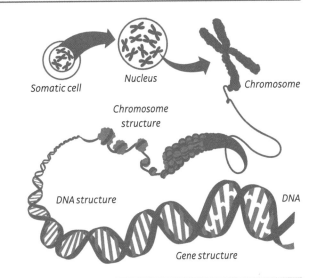

Somatic cell

Nucleus

Chromosome

Chromosome structure

DNA structure

Gene structure

DNA

MINIMAL GENOMES

Even the simplest organisms usually have **thousands of genes**, adding up to very long stretches of DNA. Compiling synthetic versions of this would be hugely difficult, so researchers tried to find the **lowest number of genes possible that would still produce a** viable, self-replicating organism: the minimal genome.

SYNTHIA

In 2010 genetic engineer **Craig Venter** and colleagues printed out their own version of the **genome of a bacterium** with only a few hundred genes, **adding extra code** (including their own names). They inserted the synthetic DNA into a bacterium from which the genome had been removed, creating what they called "**JCVI-syn 1.0**," dubbed "**Synthia**" by the media.

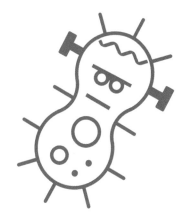

BOW AND ARROW

One of the earliest power amplifiers constructed by humankind, the bow and arrow shows the engineering prowess of prehistoric peoples.

BOW BASICS

In its simplest form, a bow is a single length of **springy**, **resilient material**, with a central **grip**, and upper and lower **limbs** on either side. It has a back and a belly, and nocks at either end for fixing the bowstring. The distance between the bow and string in a strung bow is the fistmele, or **bracing height**.

POWER AMPLIFIER

A bow is a kind of **spring** that allows the **work done** by the drawer to be **converted into and stored as mechanical or elastic potential energy** in the bent bow, which is then converted into the **kinetic energy** of the arrow when the string is released. **Energy** is **put into** the bow **slowly**, then **transferred very fast** into the arrow. Such a device is known as a power amplifier.

SELF BOWS

The earliest bows were made of single pieces of wood, and are known as **self bows**. Bowmakers learned to choose strong, elastic woods such as **yew**, **elm**, **oak**, and **ash**, and to cut them so that the more **flexible sapwood** lay on the face of the bow while the tougher, more **compressible heartwood** lay on the belly of the bow.

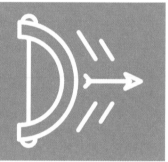

COMPOSITE BOWS

The **kinetic energy** of the arrow depends on its **speed**, which in turn depends on the **force** the bow can generate when drawn (known, when expressed in weight units, as the **draw weight**). This in turn depends on its **materials engineering**. Self bows are limited by the properties of the single material used, but in **Bronze Age West Asia**, bowmakers developed the art of **composite bows**: **layering** together materials of **complementary properties** to **combine strength with elasticity**.

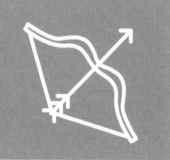

CROSSBOW

While the longbow could be devastating, it required years of training to master. Medieval engineers sought to devise a weapon of similar power that could be mastered in a day.

c.6th century BC
The ancient Greeks use giant crossbow ballistae

c.5th century BC
The ancient Chinese invent the crossbow

c.10th century AD
The crossbow is introduced to western Europe

c.14th century Invention of the steel crossbow

c.1470 Handguns make crossbows obsolete

16th century Development of sophisticated hunting crossbows

1894 Chinese forces still using repeating crossbows in the Sino-Japanese War

PARTS OF A CROSSBOW

A crossbow has a central stock or tiller, with a perpendicular bow or prod. When the string is drawn (spanned), it is held in place by a nut that can be depressed by a **trigger**. Materials used included **wood** or **metal** for the **stock**, **bone** or **ivory** for the **nut**, **hemp cord** for the **bowstring**, and for the **bow** itself, **wood**, **steel**, or a **composite of horn/whalebone**, **lathes of yew**, and **tendon**.

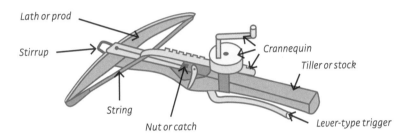

Lath or prod

Stirrup

Crannequin

Tiller or stock

String

Nut or catch

Lever-type trigger

SPANNING

The crossbow was engineered to have enormous **draw weight**, giving it **great power**. But this meant **spanning** the bow **could not be done by arm strength alone**. Techniques for spanning evolved from **bracing against the belly**, to the use of **stirrups** so that **leg strength** could be used, to the **"claw and belt" combination** of stirrup with a wire running through a belt ring that allowed the bow to be spanned by straightening the back. Larger and stronger crossbows had to be spanned with **mechanical assistance** using devices such as the **cranequin** and **windlass**.

ARMS RACE

The power of the crossbow meant that a lowly common soldier could penetrate the expensive armor of a rich knight. Knights responded by **switching from chain mail to steel plate**, but even this **could not withstand a steel crossbow shot**. Captured crossbowmen were cruelly treated by knights.

AEROSPACE & ARMAMENTS

BALLISTA

The ancient war machine known as the ballista was an extraordinary example of Roman engineering genius. Resembling a giant crossbow, it used torsion power to hurl bolts or stones.

FROM TENSION TO TORSION

The Romans got the idea for their ballista from the **Greeks**, who started off with large crossbow-like weapons that hurled stones or darts using power generated from wooden arms under **tension**. Roman ballistae, however, relied on torsion, in which energy is stored through twisting a **resistant, elastic material, and released when it unwinds**. The Romans used **animal sinew**, wound into thick cords.

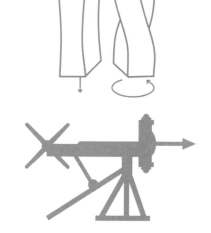

TWO ARMS GOOD

The basic plan of the Roman ballista had two **bow arms**, each one planted into a twisted skein of sinew fibers. A bowstring ran between the two arms, and, as in a crossbow, would be pulled back by a **lever** and held in place by a **pawl**. Around this was a frame with devices for **twisting the skeins** to increase **torsion**, and others, such as **winches, windlasses**, or **pulleys**, to make the lever **easier to pull back**.

SWINGING IN OR OUT?

Experts disagree over whether the arms of ballistae projected out from the skeins to give a crossbow-like appearance, or inward.

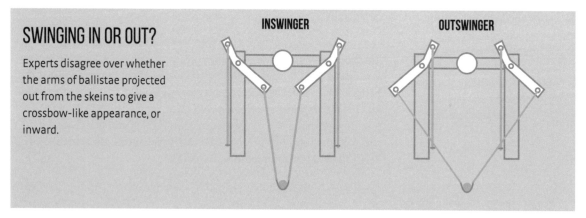

INSWINGER

OUTSWINGER

TYPES OF BALLISTA

The Roman army fielded ballistae in a **range of sizes**. The most common was the **scorpio** ("scorpion"), a small, two-man-operated ballista that fired bolts like large arrows or stones weighing 4 Roman pounds (2.8 lb). Other types included a **mobile scorpio**, known as a **carroballistae**, and **hand-held** (though probably still requiring some sort of stand) ones known as **cheiroballistae** or **manuballistae**.

POLYBOLOS

Ancient writers even described a **chain-driven, self-loading, magazine-fed ballista called a polybolos** ("multiple shot"), although it is not clear whether this was ever a practical reality.

SIEGE ENGINES

The great Bronze Age cities were protected by colossal fortifications; besieging forces needed to use all their engineering ingenuity to overcome them.

BUILD THAT WALL

As ancient cities accumulated material wealth, they sought to protect it by constructing ever larger **fortifications**. **Nineveh** in the second millennium BC was said to have **stone walls** fifty miles long, 120 feet high and thirty feet thick, while **Babylon**, in around 600 BC, was enclosed by a wall twelve miles long, 330 feet high, and wide enough for a four-horse chariot to ride on.

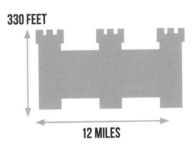

330 FEET

12 MILES

HELEPOLIS

By the third century BC, the Hellenic warlord **Demetrius Poliorcetes** commissioned a colossal siege engine named **Helepolis** ("taker of cities"), which took 3,400 men to move. According to one account, it was 135 feet high, 60 feet broad, and weighed 360,000 lb. It had **two internal staircases**; one for going up, the other for coming down.

CLOSE APPROACH

Besieging forces could try to **undermine**, **overtop**, or **break through** walls, but they always needed to get close to them, where they would be vulnerable to **arrows**, **spears**, **rocks**, and **boiling liquid**. Protection was needed. A ninth-century BC relief from the **palace of Nimrud** shows one solution: a six-wheeled **siege tower** with a central tower housing archers, and a projecting **battering ram**.

ADDED EXTRAS

Roman armies became expert at **rapid construction of siege engines in the field**. The **upper stories** might have a **drawbridge** (known as a ***pons*** or ***sambuca***) that could be dropped down on top of the wall to allow invading soldiers to stream out of the tower. The base might house a **ram**, a huge beam or log for **battering down defenses**, along with auxiliary devices to **widen breaches**, such as the ***terebus***, an iron point on the end of a pole.

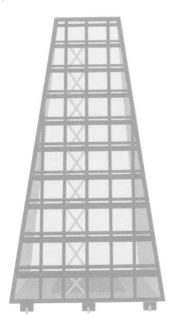

FIRE PROOFING

The main **counterattack** against siege engines was to **set fire to them**. Accordingly, they were protected with **rawhide** and **layers of rags painted with anti-inflammable substances**, or even **clad in iron plates**. They might also be fitted with **fire-suppression systems**, such as large **water-filled sacks fitted to hoses**.

TREBUCHET

A trebuchet is a catapult with a lever pivoted close to one end, powered by applying a force to the short arm.

4ᵗʰ century BC
The ancient Chinese are using traction trebuchets

6ᵗʰ century AD
The Byzantines are using petrobolos traction trebuchets

1097 Byzantine emperor Alexios I Komnenos invents the counterweight trebuchet

1199 The Trabuchus counterweight trebuchet is used at the siege of Castelnuovo Bocca d'Adda in northern Italy

c.1300 Edward I of England commissions the great trebuchet Warwolf for his campaign against the Scots

1410 Christine de Pizan's book of strategy advises the use of trebuchets alongside heavy cannon

c.1500 End of the trebuchet era

HURLING POWER

A **counterweight box** with a volume of around eighteen cubic metres could hold up to **thirty tonnes of ballast**. Such a device could hurl a 100-kilogram stone over 400 metres and a 250-kilogram stone over 160 metres. The biggest could throw **stones up to 1,500 kg**.

TRACTION, HYBRID, AND COUNTERWEIGHT

Trebuchets are classified according to the source of the force applied to the **lever**. The earliest versions were **traction trebuchets**, where the short end of the lever is pulled down via animal labor, either human or beast. **Counterweight trebuchets** use a counterweight (such as a rock-filled box), attached to the short end, to pull it down by **gravity**. **Hybrids** use a **counterweight to assist traction**.

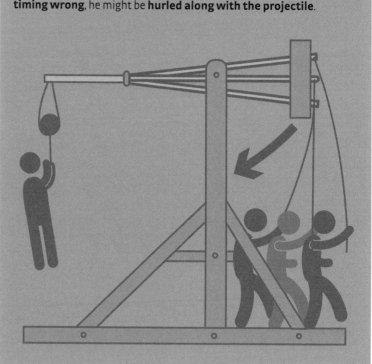

TIMING IS EVERYTHING

With smaller trebuchets, the **loader** at the end of the long arm would sometimes **add force** to the release by **hanging on to the launching end until the crucial moment**, when he would let go. If he got the **timing wrong**, he might be **hurled along with the projectile**.

GUNPOWDER

Although it has been used for a thousand years, gunpowder is still one of the most effective ways of packing a great deal of energy into a small space.

HISTORY

Gunpowder was invented by the **Chinese** before AD **1000**.

It was used for **fireworks** and **rockets**, then **guns** and **bombs**.

It was described in England in **1267** by monk **Roger Bacon**, in coded Latin.

CHEMISTRY

Black gunpowder is a mixture of **saltpeter** (75 percent), **sulfur** (10 percent), and **charcoal** (15 percent).

Saltpeter (**potassium nitrate** KNO3) provides the **oxygen**. Sulfur and charcoal burn rapidly to make gases **sulfur dioxide** (SO_2) and **carbon dioxide** (CO_2).

SULFUR

CHARCOAL

SALTPETER

USES

Fireworks use gunpowder; the **smell** is characteristic, and is partly due to the **sulfur dioxide**. The **colors** of fireworks come from small amounts of metals that are mixed with the powder—**sodium for yellow**, **barium for green**, **copper for blue**.

Modern guns use a **smokeless powder**.

Gunpowder is a **low explosive**. Modern high explosives are made from chemicals such as **dynamite** and **guncotton**.

PREPARATION AND EXECUTION

For **high efficiency**, the **ingredients** have to be **ground together** to make an **intimate mixture**.

A **teaspoonful** of gunpowder on a plate burns with a **flash**, but no **explosion**. A teaspoonful of gunpowder enclosed in a **matchbox** or a wrap of tape **explodes** when **ignited**, because the **hot gases** have to **escape**.

It's been used for hundreds of years in **guns** and by miners for **blasting rock**.

ROCKETS

Rockets are vehicles propelled by a reaction motor, which is one that employs the Newtonian principle that for any action there is an equal and opposite reaction.

REACTION PRINCIPLE

The **simplest form** of rocket is a tube, open at one end, from which matter is expelled. Following Newton's law of motion, as the matter leaves in one direction, an equal and opposite force acts in the other direction, driving the rocket forward. **The faster that fuel is expelled in one direction, the faster the rocket will travel in the other**.

TYPES OF FUEL

Rockets can operate on a **range of fuels**, from **compressed air and water** to **chemical, and even atomic, explosives**. Most rockets use fuel that burns to create **hot, expanding gases**, and it is the **expulsion of these** that **drives the rocket forward**. Rocket pioneers experimented to find the fuels that provided the **most thrust for their weight**, while also providing a **continuous burn**.

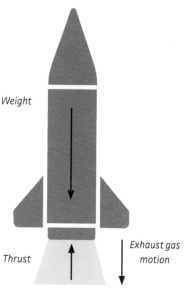

Weight

Thrust

Exhaust gas motion

JET VS ROCKET

A **jet engine** is a type of rocket engine that **expels matter wholly or partly taken in from the surrounding space**.

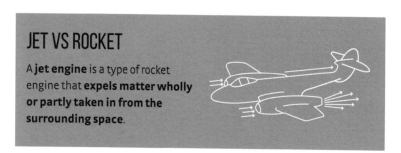

TOYS OR WEAPONS?

The **earliest gunpowder weapons** were probably **simple rockets**, created by packing **gunpowder** into the open end of a segment of **bamboo**. These could be used as **fireworks**, or as devices to spread **noise, confusion, fire**, and potentially **shrapnel**.

CONGREVES

In the imperial era, the **British military** encountered **Chinese and Indian forces equipped with rockets**, prompting Englishman **Sir William Congreve** to develop **"Congreve" rockets**. These were used by British and US armies in the early nineteenth century, but **could not match the power, range, or accuracy of artillery**.

V-WEAPONS

After the First World War, the Germans were **forbidden to develop normal artillery weapons**, but treaty obligations **did not cover rocketry**. By the end of the Second World War, a team led by **Wernher von Braun** had developed the **V-2 rocket**, a **supersonic intercontinental ballistic missile**.

EARLY CANNONS

A cannon is a large firearm, a device that uses gunpowder to propel a projectile, while it remains in place.

- **12ᵗʰ century** AD
  Europeans encounter Arab *madfaa* gunpowder weapon

- **1326** *Pot-de-fer* featured in illustrated manuscript

- **1331** First recorded use of cannons in war

- **c.1400** Cast bronze cannon

- **c.1420** First wheeled cannon

- **c.1450** Cast-iron cannon balls

- **1453** Ottoman cannons reduce the walls of Constantinople

- **1494** Charles VIII of France takes Italy with the help of light new cannons with integrated trunnions on two-wheeled carriages

ORIGIN OF THE CANNON

Although the **Chinese** started to use **gunpowder** in weapons by the Middle Ages, they do not seem to have developed **firearms**. Europeans encountered an **Arab weapon** known as the ***madfaa***, which may have been a **flamethrower** or a version of the European ***pot-de-fer***, a bowl or vase-shaped receptacle that could **hurl an iron dart** a short distance.

BOMBARDS

To **contain the explosive force** of the gunpowder without meeting destruction, engineers built early cannons known as **bombards** using **barrel-making techniques. Hoops of iron** bound **staves of iron** to create **sturdy cylinders**, which sat on wooden stocks.

METAL MONSTERS

Iron-hoop bombards could be enormous. The **Mons Meg** gun of 1457 is thirteen feet long, weighs six tons, and could fire a **stone ball** weighing 330 lb nearly two miles. A hooped-iron bombard named **Basilica**, constructed for the **1453 assault on Constantinople**, had a bore thirty-six inches across, required 200 men and sixty oxen to move it, took an hour to load, and fired a ball weighing 1,600 lb over a mile. It fell apart after just a few shots.

DEADLY IMPROVEMENTS

Factors that constrain the **power and efficacy of cannon** include the **fit between the cannonball and the bore of the cannon**; the **strength of the material** used to make the cannon; the **quality of the gunpowder**; and the **mobility of the cannon**. These are often interrelated. **Stronger metals** allowed cannons to be made **lighter without losing their integrity**, which made them **more mobile** and **easier to aim. Better engineering** produced cannonballs that **fitted tightly into the bore** and **better gunpowder** increased the **force applied to them**. By the end of the fifteenth century, cannons were **smaller**, **lighter**, **more mobile**, and **more powerful**.

CANNONS

Early cannons were restricted to siege operations. If they were to have an impact on the battlefield as part of tactical warfare, they would need to be made much lighter and more mobile.

BORING BARRELS

Advances in engineering meant that **gun barrels could be bored**, and **shot manufactured**, with **greater precision**, which **reduced escape of gas** around the shot, which in turn meant **less powder was needed to produce the same projectile force**. This in turn meant cannons could have **thinner walls**, since they needed to withstand a smaller explosion, producing **lighter and more mobile cannons**.

STANDARD CANNONS

The advent of lighter cannons presented an opportunity that was seized by the French, specifically by **Jean Gribeauval**, who became Inspector of Artillery in 1776, and **combined the light new ordnance with a standardized, lighter system of limbers and gun carriages**. He **reduced the size of a typical gun carriage team** from a dozen horses to just six horses, a ratio that would obtain until motorization.

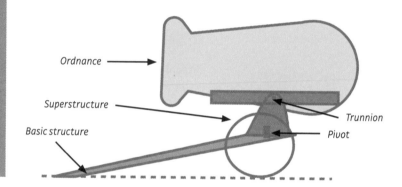

TYPES OF SHOT

Artillery could be loaded with **different types of shot** for **different targets**. **Heavy balls** were best for **smashing fortifications**, but against **personnel**, **case or grape shot** (large numbers of smaller projectiles) were more devastating. A **parabolic trajectory** can lob projectiles over walls or hills but this reduces the **velocity** of the shot, so **mortars** often use **explosive shot**.

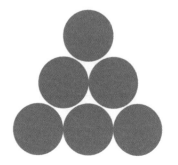

SMOOTH BORE VS RIFLING

Rifling barrels (inscribing spiral grooves on the inside) imparted **spin** to the projectiles, which could lead to **more stable, accurate flight**, but only worked with **longer, cylindrical shot or shells**. These typically had **lower muzzle velocity**, but retained **flight velocity** better than **smooth-bore shot**. Because **kinetic energy** is $\frac{1}{2}$ mass × the square of velocity, the speed of the projectile is the most important factor. Over **short distances**, smooth-bore cannons might therefore be more effective than rifled ones.

BREECH-LOADING ARTILLERY

Modern artillery is the result of over 400 years of engineering improvements to cannons, producing steel-barreled, rifled, breech-loaded cannons that are powerful but light and mobile.

BIG BERTHA

Howitzer is a class of gun that combines the **direct fire capability** of a **cannon** with the **parabolic capability** of a **mortar**. These were used to batter enemy forces and structures **from a distance**. By the **First World War**, **railway-carriage-mounted howitzers** attained monstrous dimensions. Big Bertha, the most famous railway howitzer of this time, was a German gun with a 17-in bore, which fired 1,719 lb shells over nine miles

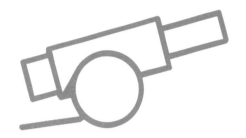

THE BREECH CHALLENGE

Muzzle-loading of cannons was slow and unwieldy, and **ruled out rifling**. **Breech-loading** would allow rifled barrels and much more **rapid fire**, but raised the problem of **obturation**: the need to **seal the edges of the breech to prevent venting of hot gas**. Solutions explored around the mid-nineteenth century included **rings of softer metal**, **"mushroom" bolts**, **interrupted screw breeches**, and **expanding shell cases**.

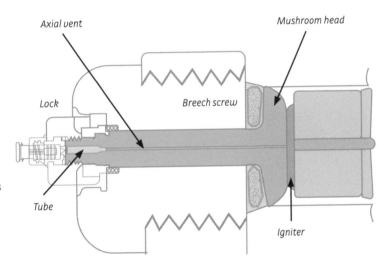

Axial vent

Mushroom head

Lock

Breech screw

Tube

Igniter

INTERRUPTED SCREW

The **interrupted screw breech** made it easy to **close a breech quickly**. The **rear of the bore** was **screw threaded**, as was the **plug used to seal it**. Instead of needing to screw the plug all the way in with repeated turns, segments of the threads on each element were removed so that the plug, mounted on hinges, could swing all the way in and then be secured with just part of a single turn, all of its threads engaging at once.

FIREARMS AND ARTILLERY

A firearm is a gun that inflicts damage on targets by launching a projectile driven by rapidly expanding high-pressure gas. Artillery is a heavier class of weapon, like a cannon.

ANATOMY

A gun has a tube called a **barrel**, closed at one end; some **gunpowder**; and a **bullet** or other projectile, which is forced out of the barrel when the gunpowder is **ignited**.

TYPES OF WEAPONRY

A **revolver** is a **handgun** with **five or six bullets** in separate **chambers** in a **revolving cylinder**, so that they can be fired in quick succession. **Samuel Colt's** revolvers were widely used in the United States in the mid-1800s, and revolvers are still popular in the US today.

Pistols are **handguns** with a **chamber that is integral** with the barrel.

The earliest guns were made by the **Chinese** in **c.ad 1000**, when they invented gunpowder. The barrels were **bamboo tubes**.

A **cannon** is a **large heavy** gun, usually **mounted** on a **carriage** or firm support; now replaced by **artillery**.

Later guns fired bullets or occasionally shot. The **blunderbuss** was a short gun with a flared barrel. It fired shot, and was effective only at **short range**, but caused much **noise and destruction**.

Artillery pieces have **barrels** ranging in diameter from a few inches to a couple of feet. They are **mounted on tripods**, **carriages**, **rails**, or **ships**, and generally fire shells containing high explosive or **incendiary** material. The propellant is usually a combination of **nitrocellulose** and **nitroglycerine**.

Rifles are long guns fired from the **shoulder**. They use single bullets, which are mounted in a **brass case** containing the propellant, usually nitrocellulose. The barrel has **helical grooves** that make the **bullet spin**, thus improving the accuracy of the weapon. Some rifles are accurate at more than **3,000 feet**.

The **modern shotgun** has long barrels, and usually fires **lead shot**, which comes in various sizes: no. 4 shot is **3.3 mm** in diameter; no. 6 shot is **2.8 mm** in diameter. The shot, or pellets, are packed with the gunpowder inside a cardboard cartridge. Shotguns are accurate to about **150 feet**.

AIRSHIPS

Airships, or dirigibles, are essentially large gasbags filled with a lifting gas that is less dense than the air surrounding them. Despite being huge lumbering beasts, engineers keep finding new ways and places to use them.

1783 The Montgolfier Brothers fly two men in a hot-air balloon, and Jacques Charles flies with a copilot under a balloon filled with hydrogen, the lightest gas of all

1785 Jean-Pierre Blanchard crosses the English Channel under a balloon equipped with flapping wings

July 1900 The major airship industry begins with the Luftschiff Zeppelin LZ1, built by Count von Zeppelin in Germany

1914–18 In the First World War the Germans use airships for reconnaissance and for bombing, but they are vulnerable to bad weather and attack by other aircraft

1930s German Zeppelins carry passengers in amazing comfort, with private cabins, viewing platforms, and dining salons. They are faster than ocean liners, and can carry more passengers than conventional aircraft

THE HINDENBURG DISASTER

In a **major disaster** on **May 6, 1937,** the **Hindenburg caught fire and crashed** on its approach to Lakehurst, New Jersey; **thirty-six people died**, although, remarkably, **sixty-one survived**.

THEN AND NOW

Early airships were just big balloons. Passengers or freight were carried in **gondolas slung underneath**. **Helium-filled** airships are used today for **surveillance (spying)**, for **exploration**, and for **carrying heavy loads to remote places**, such as dams high in the mountains, where there are no roads.

HYDROGEN VS HELIUM

Although **hydrogen is the lightest** (least dense) gas, and is readily available, **it is highly flammable**, and this always posed a **risk for airships that used hydrogen** for lift.

The **Hindenburg disaster marked the end of passenger-carrying airships filled with hydrogen**. **Helium is denser than hydrogen**, and **supplies are scarce**, but it has the advantage that **it will not burn**.

POWERED FLIGHT

Powered flight was one of the greatest engineering challenges of the industrial age, requiring a vehicle that could generate and sustain enough lift to overcome its own weight.

INCLINED PLANES

Some early aviation pioneers tried to **copy birds** by designing **flapping wings**. It was Englishman **George Cayley** (1773–1857) who developed the concept of a **fixed wing**, in which a plane surface is set at an angle to air flowing across it. As air is **deflected down** from the wing, an equal and opposite force will **push the wing up**.

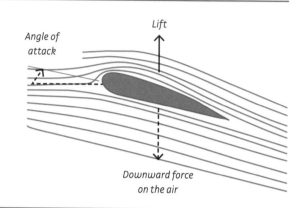

Lift

Angle of attack

Downward force on the air

VECTOR FORCES

Cayley was also the first to identify the **four vector forces** that act on an aircraft: **thrust**, **lift**, **drag**, and **gravity**.

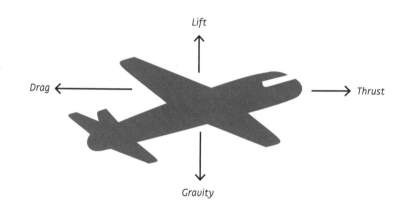

Lift

Drag

Thrust

Gravity

THE AERIAL TRANSIT COMPANY

Cayley's ideas were taken up by English engineers **William Henson** and **John Stringfellow**, who developed a **monoplane design**, where a **large inclined plane wing** would generate **lift**, a **tail fin** and **rudder** would control **pitch** and **yaw**, and a **lightweight steam engine** would drive **propellers** to generate **thrust**. In 1843 they even tried to found an international airline, the **Aerial Transit Company**.

STRINGFELLOW'S TRIPLANE

Their venture failed and Henson moved to the United States, but Stringfellow persisted. Although his steam engines were light, they still could not provide a high enough **power-to-weight ratio** to lift a human passenger, but a **triplane model** version of his aircraft successfully achieved **tethered powered flight** at the **Aeronautical Exhibition** held at the **Crystal Palace** in 1868.

MACHINE GUNS

A machine gun is a fully automatic firearm designed to fire rifle cartridges in rapid succession for sustained fire: a purpose that transformed battlefield tactics, with horrific results.

FULLY AUTOMATIC

The basic principle of a machine gun is to **automate** the processes of **firing a bullet**, **expelling the empty shell case**, **chambering the next round**, and **firing again**. To do this, some form of **power or drive is needed**. According to legend, French-American inventor **Hiram Maxim**, who invented the **first fully automatic machine gun** in 1885, was inspired to do so when he fired a rifle and the **recoil** hurt his shoulder. Maxim realized that he could use the recoil to drive the machine gun.

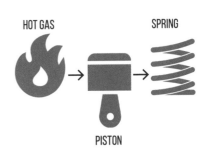

GAS OPERATION

Recoil—the equal and opposite reaction to the launch of a projectile, predicted by **Newtonian laws of force**—is just one of the possible sources of drive for a machine gun. Another is the **pressure of the hot gases** created when the **powder charge explodes**; in the **Lewis machine gun**, for instance, this gas is used to drive a **piston** attached to the **bolt**, drawing it back, while at the same time **compressing a spring** that will **rebound and close the bolt**.

HOT GAS   SPRING

PISTON

KILLING POWER

While even the most expert rifleman could fire at most fifteen shots in a minute, a machine gunner could fire 600 shots a minute.

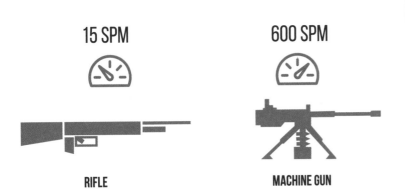

15 SPM   **600 SPM**

RIFLE   **MACHINE GUN**

BROWNING HEAVY MACHINE GUN

US gun designer **John Browning** designed a **recoil-operated heavy machine gun** that was **lighter**, **simpler**, **cheaper**, and **more reliable** than Maxim's. When he demonstrated it to the US Army in May 1917, he fired off 40,000 rounds with a single minor component failure.

THE WRIGHT BROTHERS

The Wright brothers would succeed where so many others had failed because of their ingenuity and methodical approach to aviation engineering.

BICYCLE BROS

Wilbur and Orville Wright were **bicycle engineers** who had nurtured an interest in aviation ever since their father bought them a toy helicopter in 1878.

STEP BY STEP

In developing their heavier-than-air powered flyer, they progressed methodically through **prototyping** and **testing** stages, starting with a **kite**, proceeding to a **glider**, and finally making a **powered flyer**.

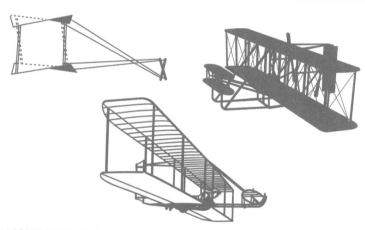

WING WARPING

Inspired by their observations of **birds**, the Wright brothers designed **"warping" wings** that could be **angled for balance and control**.

LIGHT WEIGHTS

They developed a **lightweight "floating" wooden structure**, with **spars sitting in sewn-in pockets in the fabric**. Making the fabric an **integral part** of the design **saved weight** and made the structure **more resilient**.

ALUMINUM ENGINE

The brothers **developed their own engine**, which included a **revolutionary aluminum casing**. Extremely **light**, aluminum would become a crucial material in aviation.

AIRBORNE

On December 17, 1903, with **Wilbur piloting**, the Flyer achieved the **first free, controlled flight of a powered, heavier-than-air plane**. It flew 852 feet in 59 seconds.

TANKS

An engineering solution to the horrific challenges of the First World War battlefield, the tank would change the nature of warfare.

1485 Leonardo da Vinci designs a conical armored chariot

1801 Invention of caterpillar track

c.1900 German-English engineer F. R. Simms designs an armored quad bike and car

1903 H. G. Wells predicts tanks in his story "The Land Ironclads"

1904 French firm Charron, Girardot and Voight builds armored motor cars

1914 Holt motor tractors in use by the British army

1915 The British government sets up the Landship Committee to pursue the development of tanks

1916 Tank Mk I commissioned; first battlefield deployment in September

BATTLEFIELD CHALLENGES

The battlefields of the Western Front in the First World War posed terrible challenges for **infantry**: withering **machine-gun fire**; broken, impassably muddy **terrain**; **antipersonnel barriers** such as **barbed wire**; and **entrenched defensive positions**. The balance between defense and offense had been tilted dramatically in favor of the former.

LAND IRONCLADS

H. G. Wells's 1903 story "The Land Ironclads" described huge armored **"landships"** with caterpillar-like treads, which overran normal battlefield defenses. The story suggested elements needed to overcome the battlefield challenges: **engines** that could generate enough power to move heavy armor, and a **locomotive system** that could cope with **battlefield terrain**.

MOTOR TRACTORS

Some of these innovations had already been combined in an existing vehicle: the **Holt motor tractor**—a large, heavy, powerful vehicle with **caterpillar treads** to cross broken ground. British engineers used this model as the inspiration for a new class of military vehicle: a **mobile gun platform** with a **low center of gravity**, **heavy armor**, and **caterpillar tracks**.

MK I TANK

The Mark I tank was rhomboidal in shape, with a caterpillar track running all the way round the hull. The angled thrust of its hull helped it to negotiate obstacles and meant the tracks were less likely to fall off.

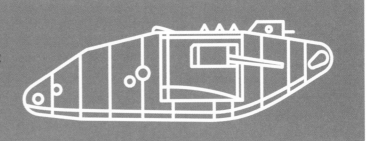

UAV DRONES

In common parlance, "drone" is short-hand for a remote-controlled vehicle, typically a pilotless aircraft usually known as an unmanned aerial vehicle (UAV).

1849 Bomb-bearing pilotless balloons launched by the Austrians during an assault on Venice

1898 Serbian inventor Nikola Tesla demonstrates radio-controlled unmanned boats

1915 American inventor Elmer Sperry designs the aerial torpedo, a lightweight airframe packed with explosives and stabilized with a gyroscopic guidance mechanism

1939 Radio remote-control toy aircraft enthusiast (and British film actor) Reginald Denny sells thousands of OQ-2 Radioplanes to the US military

1940s Radio-controlled planes include the German Argus As 292, British Queen Bee, and American TDR

1960s–70s US military use the AQM-34 Ryan Firebee drone for reconnaissance

1970s and 80s Glider-style drones developed by the Israelis

1994 Predator drone's first flight

TELAUTOMATONS

The key enabling technology for **remote control** was **wireless radio transmission**. One of the pioneers of **radio technology** was the Serbian inventor **Nikola Tesla**, who incorporated **sophisticated encoded radio control** into a **"Telautomaton" electric motor drone boat system** that he demonstrated at the **Electrical Exposition** in **Madison Square Gardens** in 1898.

BREAKING THE LINE

Radio allowed remote control, but only so long as operators maintained a **line-of-sight** contact with the vehicle. Only when **live-feed cameras** became available was it possible to break the line of sight.

UAV LOGIC

Compared to conventional aircraft, UAVs are **cheaper**, more **expendable, easier to transport and deploy**, and **use less fuel**. Above all, they **keep operating personnel out of harm's way**.

OUTNUMBERED

In 2005, just 5 percent of US military aircraft were drones; **today UAVs outnumber manned aircraft in the US military**.

JET ENGINE

There was a limit to the speeds that could be achieved by internal combustion engines driving propellers. A new engine design using reaction principles promised a massive boost in power, if the tricky engineering could be accomplished.

REACTION ENGINES

A jet is a reaction engine that generates **thrust in one direction by expelling a fluid medium in the other direction**. Aviation jets work by **taking in air at the front of the engine, compressing it, and burning it with fuel to generate expanding hot gases** to expel at the other end.

TURBOJETS

The first jet engines were **turbojets**, which use the **expanding combustion gases** to **spin a turbine** before **exhausting through the rear nozzle**. The turbine powers the **compressor** at the front of the engine that **compresses the incoming air**.

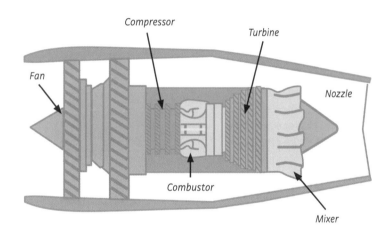

Compressor

Turbine

Fan

Nozzle

Combustor

Mixer

SECOND WORLD WAR JETS

British pilot **Frank Whittle** patented the **first turbojet design** in 1930, but it took him another eleven years to get one to fly. In 1939 German engineer **Hans von Ohain** succeeded in developing the **first working turbojet**, which powered a **Heinkel He 178**. In 1942 **General Electric** built **the first US jet plane**, the **XP-59A experimental aircraft**.

TURBOFANS

Most jet engines today are **turbofans**, which use the **turbine to power a fan at the front of the engine**, before the compressor. The fan **draws in extra air** and **splits the airflow** so that the **extra air flows around the outside of the engine before mixing with the hot exhausts at the rear**. This increases thrust without **increasing fuel consumption**, and also **makes the engine quieter**.

HELICOPTERS

*A fixed-wing aircraft generates lift by moving the whole vehicle forward through the air.
A helicopter works on the principle that if only the wing travels through
the air, the aircraft as a whole can rise vertically.*

DA VINCI'S HELICOPTER

Several **trees** produce **spinning or rotating seeds**, which drop more slowly and so can be dispersed more widely by the wind. Perhaps inspired by this natural design, both the **ancient Chinese** and **Leonardo da Vinci** designed **concept vehicles lifted by rotating planes**. In da Vinci's case it was a **helical screw**.

LIFTING BLADES

The **blades of a helicopter's rotor** are just **like airplane wings**, with a **cross section that generates lift when air moves over them**—or when they move through the air.

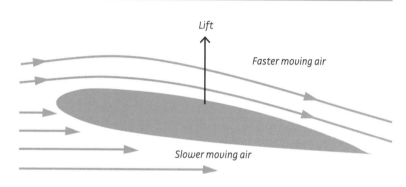

Lift

Faster moving air

Slower moving air

TORQUE TALK

A **single rotor on its own** exerts a **rotational force (torque)** on the **body of the helicopter** even as it spins. **To stop the body spinning in the opposite direction to the rotor**, helicopter designers used two different methods:

Counter-rotating blades: two rotors spinning in opposite directions cancel out each other's torque.

Tail rotors: Spinning in the vertical plane, this rotor **counters the body torque** and can also be **used for control**.

SIKORSKY

Although not the first to design or even fly a helicopter, Russian-American aviation engineer **Igor Sikorsky** is known as the **father of helicopters** because his **VS-300**, which lifted off in 1939, was the **first successful design**.

BOUNCING BOMB

British aircraft designer Barnes Wallis was convinced that engineering would play a decisive role in the outcome of the Second World War. With his "bouncing bomb," he got the chance to show how.

UNBUSTABLE

Dams in Germany's **Ruhr valley**, which supplied water and power to industry, were obvious targets for **Allied bombing raids**, but they were simply too big to be damaged by bombs **unless the charges detonated right up against them. Nets protected them from torpedoes.** How could a **mine** be delivered to the **base of such a dam?**

UPKEEP AND HIGHBALL

Barnes Wallis had the idea for some sort of missile that would **skip across the surface of the water, jumping over the anti-torpedo nets**, before clattering into the dam, and falling back down its face to explode at the waterline. He developed a small version, codenamed **Highball**, for **deployment against battleships**, and a large one, **Upkeep**, to **launch against dams**.

BACKSPIN

The bombs (actually bouncing mines) **needed to spin backward at exactly the right rate** (500 rpm) to **make sure that they would roll back down the face of the dam** when they hit. The backspin also **ensured the bomb would fall behind the path of the plane to give the bombers more time to get clear**.

BOUNCING BARRELS

The original design was for **spherical bombs**, based on Wallis's **original experiments** with **marbles in a bathtub**, but a **barrel shape** was eventually selected.

DAMBUSTERS

The dams had to be hit after the spring rains had filled the reservoirs, and in the space of just two months a squadron of specially adapted **Lancaster bombers** was trained in the **difficult and dangerous low-flying approach needed to launch Upkeep bombs**. On May 16–17, 1943, the 617 Squadron **successfully breached two massive dams and damaged another**. Of the nineteen planes, eight were lost and **fifty-three men lost their lives**.

| 19 PLANES | 8 LOST | 53 KILLED |

ATOM BOMB

The atomic bomb is usually thought of as a great scientific achievement, but in fact the science behind it was known since before the war—the true challenge was an engineering one, on a colossal scale.

CHAIN REACTION

The key scientific insights behind the atom bomb included **Einstein's famous equation E=mc²**, which showed that a **tiny amount of mass could be converted into a huge amount of energy**, and Hungarian physicist **Leo Szilard's theory that a chain reaction could be set off in an element with the right kind of unstable radioactive nucleus**—a so-called "**fissile material**."

FISSILE ISOTOPE

In 1940, the US government set up the **Uranium Project** to look into the possibility of an atom bomb. It concluded that the fissile material needed was a particular **isotope of uranium, U-235**, but U-235 **accounts for just 0.7 percent of naturally occurring uranium**. Separating out enough for a bomb would be an enormous engineering challenge.

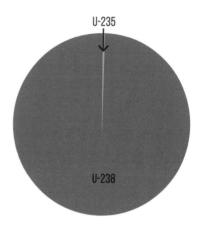

NATURALLY OCCURRING URANIUM

URANIUM ENRICHMENT

The industrial-scale effort to separate uranium isotopes and produce enough U-235 for a bomb resulted in the construction of vast plants at **Oak Ridge** in **Tennessee** and **Hanford** in **Washington**. Techniques employed included **thermal diffusion**, **electromagnetic separation**, and **gaseous diffusion**.

HIROSHIMA

On August 6, 1945, Little Boy exploded 1,900 ft. above Hiroshima. Despite measuring just ten ft. by twenty-eight inches, and containing just 140 lbs of uranium fuel, it had a **force equivalent to 15,000 tons of TNT**. The **shock wave** at ground zero generated **winds of 980 mph** and **pressure equivalent to 8,600 lb per square feet**. Even a third of a mile away, the wind speed was 620 mph. **Around five square miles of the city was flattened**.

BOMB DESIGN

At **Los Alamos** in **New Mexico**, a team of scientists worked to design the actual bomb. The bomb that would be dropped on **Hiroshima**, nicknamed **Little Boy**, was a **"gun" type bomb**, in which a **pellet of U-235 of subcritical mass** was fired into a receiver of U-235, also of subcritical mass; when **combined**, they **reached critical mass** and set off an **uncontrolled nuclear fission chain reaction**.

SPUTNIK

The launch of the first artificial satellite on October 4, 1957, inaugurated the Space Age and set in motion the space race between the USSR and the US.

NEWTON'S CANNON

The concept of a satellite dated back to the work of **Isaac Newton** in the seventeenth century. Building on the **orbital mechanics** of **Kepler**, Newton showed how **gravity caused a cannonball fired horizontally to fall to Earth in a parabolic trajectory**. He then asked, what if the cannonball were fired with such **velocity** that its **parabolic trajectory carried it just beyond the curvature of the Earth?** It would be perpetually falling to Earth but perpetually missing: in other words, it would **orbit around the planet** much like our natural satellite, **the Moon**.

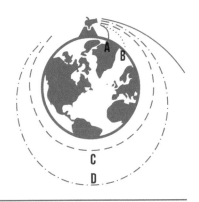

OBJECT D

The success of the **German V-2 project** showed that **long-range ballistic rocket technology already existed**. Rocket visionaries such as the Russian scientist **Konstantin Tsiolkovsky** had predicted since the early 1900s that rockets—perhaps in a **"train" of multiple stages**—could propel travelers into space. Russian rocket scientists suggested an **unmanned object could be launched into orbit as an interim stage**, and development began on an **ambitious satellite packed full of scientific instruments**, known as Object D.

SIMPLEST SATELLITE

The **Russians were determined to beat the United States into space**, and when it became apparent that their **R7 rocket** would **not** be **powerful enough to lift Object D**, they switched to plan B: a **"simplest satellite"** or ***prosteishy sputnik***. Codenamed PS, this would become known to the world as **Sputnik 1**.

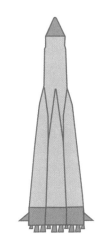

SPUTNIK DESIGN

Sputnik was formed from **two insulated, polished hemispherical shells**, enclosing a **radio transmitter**, **batteries**, **a fan for regulating internal temperature**, and **switches that would regulate the frequency of the radio transmission in response to heat and pressure changes**.

APOLLO PROGRAM

The defining engineering achievement of the Space Age, the moon landings were the culmination of the Apollo rocket program.

TEAM EFFORT

The Apollo space program involved the work of **400,000 engineers, scientists, and technicians**. **Over 20,000 companies and universities** worked on the moonshot. Countless engineering challenges were overcome, including the design of the **fundamental mission architecture**, the development of the **most powerful rocket in history, anticipating conditions in the vacuum of space**, and **designing a light but maneuverable lunar landing craft**.

BIG BOOSTER

The **Saturn V rocket** developed to launch the moonshot would be far larger than any previous rocket. It had **three stages made up of over 3 million parts, stood 363 feet tall, and produced 7.5 million pounds of thrust**.

MISSION ARCHITECTURE

A vital early decision was **how to land the astronauts on the Moon**. Thanks to the persistence and vision of **NASA engineer John Houbolt**, the **Lunar Orbit Rendezvous model** was adopted. It was a complex and technically demanding plan to use a **single launch vehicle** to send a **mother ship and landing craft directly to the Moon**, but crucially it **meant the spacecraft could be smaller**.

VACUUM PACKING

Engineers had to consider how components would respond to the **harsh conditions of space**. For example, they **identified that supercooled fuel supply lines** that on Earth were protected by a layer of frozen water vapor **would be vulnerable in a vacuum**, and **added simple steel mesh reinforcement to prevent disaster**.

STANDING ROOM ONLY

Another classic engineering challenge was **reducing the weight of the Lunar Module**. Seated astronauts needed **large, heavy windows to ensure an adequate field of view**; engineers realized that **if the astronauts simply stood up, they could make do with smaller, lighter windows**.

GLOBAL POSITIONING SYSTEM

One of several global satellite navigation systems, the global positioning system or GPS has become synonymous with this extraordinary technology.

GNSS

The Global Navigation Satellite System (GNSS) actually **comprises a number of different systems**, each operating with their own constellation of satellites. As well as the **US NAVSTAR GPS**, there are the **Russian GLONASS** and **EU Galileo satellite systems**, with a **Chinese version, BeiDou-2**, on the way.

SATELLITE SKY

The GPS system relies on **at least twenty-four satellites in high Earth orbit** (c.12,500 miles up), each one traveling at around 8,700 mph and orbiting the Earth every twelve hours. The satellites are arranged so that there is **a line of sight to at least four of them from any point on Earth at any time**. Each satellite carries an **incredibly accurate atomic clock**.

TRILATERATION

GPS receivers work out their position on Earth by **picking up time signals from the satellites** (usually from six to twelve of them) and **using these to determine the positions of the satellites**, which in turn shows the receiver's position by **trilateration**, which is a method of **determining positions using the points of intersection of three overlapping circles or spheres**.

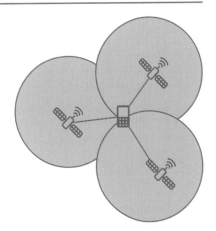

RELATIVISTIC CLOCKS

The **altitude** and **speed** of the satellites means that **relativistic effects have to be taken into account** for earthbound receivers **to remain synchronized with the clocks on the satellites. Failure to do so would cause the system to accumulate positioning errors of over six miles a day**. One clever engineering solution is that the **clocks on the satellites have been set to run more slowly than Earth clocks**.

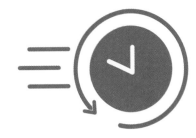

HUBBLE SPACE TELESCOPE

Launched in 1990, the Hubble Space Telescope (HST) was an audacious solution to the biggest problem facing earthbound telescopes, but it would need an engineering rescue mission before it became fully functional.

ATMOSPHERIC DISTORTION

Ground-based telescopes have to look at the stars through the **Earth's atmosphere**, and the **gas and dust absorb light and cause optical interference that limits resolution**. **Atmospheric distortion** from **shifting air packets** is why **stars seem to twinkle**. One solution is to **site telescopes on mountaintops in arid regions**, but an even better solution is to **put the telescope in space**.

LIGHT GATHERING

Hubble orbits 353 miles above the surface of Earth, and uses a 94.5-inch-wide primary mirror to **gather much more light than it would be able to do on Earth.**

CASSEGRAIN REFLECTOR

The HST is a type of telescope known as a **Cassegrain reflector**, in which a **primary mirror gathers light and reflects it onto a secondary mirror that focuses and reflects it toward the detector instruments**. Because these are behind the primary mirror, the secondary mirror **reflects the light through a tiny hole in the center of the primary.**

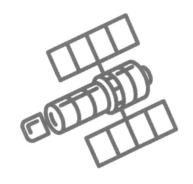

OUT OF FOCUS

Soon after the HST was deployed, it was apparent that it had a **severe problem**. An **unnoticed flaw in the mirror**, known as **spherical aberration**, meant that it was **not focusing light properly** on the detectors. Fortunately, the HST had been **specifically designed to be repairable and upgradable**. In 1993, **astronauts** from the **Space Shuttle** installed **a fix** and since then multiple other missions have seen the **detectors upgraded and replaced**, along with **other maintenance that greatly extended the lifespan of the telescope.**

INTERNATIONAL SPACE STATION

The largest space station ever constructed, the International Space Station (ISS) is the result of a truly international engineering collaboration.

MULTIPURPOSE

The construction of the ISS needed to provide **living space**, **life support**, **work space**, and **scientific equipment**; **solar arrays for power generation**; and **docking connections** to allow crew and cargo spaceship docking, and expansion by addition of new modules.

PRIME REAL ESTATE

The ISS is big. It is **as large as an American football field**, and has **more living and working space than a six-bedroom house**. Amenities include **six sleeping quarters**, **two bathrooms**, a **gym**, and a **360-degree-view bay window**. It has **fifteen pressurized modules**, including **laboratories**, **living quarters**, **docking compartments**, **airlocks**, and **nodes**, and **can accommodate seven astronauts**.

POWER AND WATER

Large **arrays of solar panels**, covering **more than half an acre** in total, supply **eighty-four kilowatts of power**. A **water-recovery system** helps crew members **recycle water**, but the station still requires a **third of a gallon a day of water supply**, which must be **sent up from Earth**.

MOBILE HOME

The **ISS orbits the Earth every ninety minutes**, traveling at five miles/second. **Every day it travels sixteen times around the Earth**, passing through **sixteen sunrises and sixteen sunsets**.

UNSUNG HEROES

Although the crew and science modules get all the glory, arguably the most important elements of the ISS are its least glamorous. One is the **fifty-five-foot robotic arm/ crane**, known as **Canadarm2**, which has **seven joints**, **two "hands" (a.k.a. end-effectors)**, and a **125-ton payload capacity**, and is used to move **modules**, **experiments**, and even **astronauts**. The other is the **Integrated Truss Structure**, the **backbone of the ISS**, on which are mounted most of the **solar arrays**.

ISS statistics	
Mass	861,804 lb
Length	240 ft
Width (along truss, arrays extended)	356 ft
Height (nadir–zenith, arrays forward–aft)	66 ft
Habitable volume	13,696 ft³

FUTURE WEAPONS

Since the first hominins fashioned stone weapons, engineering prowess has been the primary determinant of military power. The weapons of tomorrow are being engineered today.

DIRECTED ENERGY WEAPONS

Since the days of early science fiction, energy weapons such as **laser** and **ray guns** have been the dream of military engineers. But the engineering challenges involved remain daunting. **The biggest problem is power**: an **effective laser needs a power plant the size of a truck**. Even if future technology shrinks them to human-portable size, their **extreme energy density would make them more effective as hand grenades than sidearm power packs**.

HYPERSONIC PROJECTILES

Moving too fast to be intercepted, **hypersonic projectiles** could give a vital offensive advantage. But the challenges involved include overcoming **extreme heating and predicting the flow characteristics of air at hypersonic speeds**. One solution is to **equip projectiles with protective covers or capsules**, known as **sabots**, that **drop away after firing**.

NEW WEAPONS, OLD PROBLEMS

Cutting-edge technology such as **lasers** and **railguns** (which use **magnetic levitation to accelerate projectiles to very high levels of kinetic energy**) faces old-school challenges, such as **overheating** and **wear and tear of gun barrels. Current railgun launchers**, for instance, **can fire only a few projectiles before failure**.

SMART BOMBS

Precision-guided munitions, a.k.a. smart bombs, have made **artillery and bombing much more accurate** and thus politically palatable, but they have also **driven up costs**. Military engineers are now looking to **save money** by developing **simple, robust, transferable guidance mechanisms** that can be **retrofitted to existing ordnance**.

ELON MUSK

The most exciting and visionary engineer of the age, South African–born Elon Musk is fearless in his quest to disrupt traditional industries, with transformative results.

NO FEAR

Musk takes on immense industrial and commercial challenges, confident that clever application of engineering can make **radically different approaches viable and cheap**. This philosophy has allowed him to disrupt industries such as **car making**, **spaceflight**, and **tunnel boring**.

ELECTRIC DREAMS

Musk believes that essential technologies to head off **climate change** include **rapid electrification of the transport sector** and a **massive increase in battery storage and renewable-energy generation**. To make **electric cars** a reality he determined to build an **attractive electric car** with a **long battery range**. Starting with an **initial high-end model to fund his expansion**, he has successfully made **Tesla Motors** a **mass-market car manufacturer** that many **traditional auto giants are now seeking to copy**.

GIGAFACTORIES

Affordability is one of the keys to Musk's dreams of realizing **disruptive breakthrough technologies** such as **electric vehicles**. This in turn depends on **scaling his operations** to achieve **economies**, and to this end he has pushed the **development of colossal installations** he calls **gigafactories**, to **bring down the unit price of batteries and cars**.

REUSABLE SPACESHIPS

To bring down the high cost of **space launches**, Musk determined to make his **spaceships reusable**, pioneering ambitious **soft-landing technology**. His companies also use techniques such as **rapid prototyping** and **aggressive testing** to make rapid engineering gains.

MISSION TO MARS

To realize his dream of **Martian colonization**, Musk has developed a **mission architecture** involving a **rapid**, **cheap launch cadence of reusable spaceships** to get significant mass from Earth to orbit, thus enabling **orbital construction of a Mars-capable spaceship or fleet of ships**.

SPACE ELEVATOR

The ultimate engineering solution to the problem of the expense of escaping Earth's gravity, the space elevator seeks to marry simple principles with science-fiction materials.

TETHERBALL

The idea of the **space elevator** is to establish a **permanent bridge from the Earth's surface to orbit**, using the same principle at work in a tetherball game, where a **ball on the end of a string whizzes around the top of a post**. If the ball were in outer space and the other end of the string were on Earth, the **orbital motion of the ball would hold the string taught**. The space elevator concept is to **climb up and down this taut "string" with a traveling platform** of some sort.

COUNTERWEIGHT AND CABLE

The elevator would consist of a **counterweight** launched into space that would eventually sit in a **geosynchronous orbit above the equator**, some 60,000 miles up, with a **cable** stretching all the way down to an **anchor point** on the equator—probably in the **middle of the ocean**. Some kind of **platform** would crawl up and down the cable, heading for a **space station** sited in orbit.

CHALLENGES

The biggest obstacles to the concept of the space elevator include the **material needed for the main cable** and the problem of **space debris severing the thin cable**.

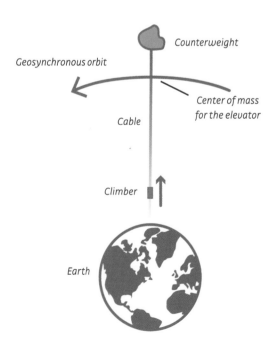

Counterweight

Geosynchronous orbit

Center of mass for the elevator

Cable

Climber

Earth

CARBON NANOTUBE RIBBON

The material needed for the main cable of the elevator would have to be **incredibly strong** yet also **light**, or else its **own weight would become too colossal**. Such a material does not presently exist, but it is hoped that in the future it may be possible to make a ribbon from **carbon nanotubes**—or perhaps a **braid of single nanotubes stretching the entire distance**.

DYSON SPHERE

What are the limits of engineering? How far can engineers go? What wonders could be achieved by an advanced civilization? Could they reengineer their entire solar system?

WASTED ENERGY

The Sun **radiates** vast quantities of **energy into space**, only a fraction of which makes it to the Earth's surface. In 1960 visionary engineer **Freeman Dyson** proposed that an advanced civilization seeking to utilize all of the solar energy produced by its sun might take **solar power** to its logical conclusion and entirely **encircle a star with solar-power-gathering or solar-power-reflecting material**, a megastructure known as a **Dyson sphere or shell**.

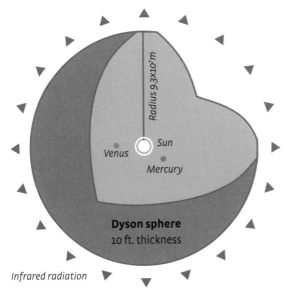

Radius 9.3x10³ m

Venus
Sun
Mercury

Dyson sphere
10 ft. thickness

Infrared radiation

DYSON RINGS

Dyson himself recognized that an **actual solid shell of material** would probably be **physically impossible to engineer** because of **tidal stresses**. He proposed that **a virtual shell** of **independently orbiting solar arrays** should be built up, probably starting with **rings encircling the star**, which would gradually **multiply until they completely blanketed the star**.

ET MARKER

Dyson proposed the sphere concept as a way to look for the **light signature of an advanced alien civilization**. He argued that any civilization that lasted long enough would eventually create such a **megastructure to meet its energy needs**, and that telescopes could search for the **tell-tale infrared signature of a Dyson sphere**, thus **detecting the presence of alien super-engineering**.

CANNIBALIZING PLANETS

One suggestion is that material for construction of a Dyson sphere might come from **disassembly of an entire moon or planet**. Future humans, for instance, might **cannibalize Mercury for iron and oxygen** to create **a layer of reflective hematite**.

ELECTROSTATIC GENERATORS

Engineering was key to the earliest scientific inquiries into the existence and nature of electricity, with baroque machines created to generate increasingly powerful electrostatic charges.

AMBER RUB

Electrical phenomena in the **ancient world** were associated with the **static electric effects** that could be generated by **rubbing amber** (fossilized pine resin), from the Greek name for which—***electrum***—seventeenth-century philosophers derived the term "**electric.**" Amber charged in this way might **glow** and **emit sparks**.

LITTLE GLOBES

German inventor **Otto von Guericke** created the **first custom-built electrostatic generator** by creating a **globe of sulfur** that **sat in a wooden cup** and **could be spun. Rubbing it by hand** generated **static charge**, which could be used to **draw sparks** or **give shocks**.

BEER-GLASS GENERATORS

Later it was discovered that **glass globes** worked as well as sulfur, and **Johann Heinrich Winckler** devised a **pedal-operated spinning machine** that **revolved a glass object against a rubbing surface** to generate **static electricity**. Winckler found **beer glasses** to be the ideal objects.

ELECTRIC KISSES

The static charges generated by such machines were used for **entertainment as well as research**. A **popular salon pastime in eighteenth-century France** was the "**electric kiss**," where a young lady would be **charged with static from a generator** and **then discharge it**, with accompanying tingle, when kissed.

LEYDEN JAR

One of the earliest products of electrical engineering was
an early form of capacitor, called the Leyden jar.

ELECTRIC FLUIDS

In the eighteenth century, scientists were still **groping for an understanding of electricity**. It was **believed to be some sort of invisible fluid or vapor, known as effluvium**.

FLUID STORAGE

At least two scientists in Europe independently concluded that the proper place to store this "electric fluid" would be in a bottle. **Charge from an electrostatic generator was fed via a wire into a bottle partially filled with water.** When **Petrus van Musschenbroek**, of the University of Leyden in the Netherlands, **charged a bottle in this fashion** and then went to **disconnect the charging wire**, he **received** such a **shock** that he told a colleague he would not repeat the experience for the whole kingdom of France.

BETTER JARS

Further experimentation showed that the jars **functioned better without water**, and even better when **lined with metallic foil** on either side of the glass.

FRANKLIN'S SQUARE

Benjamin Franklin simplified the jar to a **pane of glass with foil on either side**, known as a **Franklin square**. Later versions **eliminated even the glass**.

CAPACITORS TODAY

Modern capacitors—devices for **storing electrical charge**—consist of **two charged plates separated by an insulating material**. In principle, therefore, they are just the same as the old Leyden jars or Franklin squares, with conductive metal foil separated by insulating glass.

DANGER! HIGH VOLTAGE!

Leyden jars can indeed be **lethally dangerous**; a jar with a capacity of just an eighth of a gallon can give a **fatal shock**.

VOLTAIC PILE

A landmark bit of electrical engineering turned a curious phenomenon into the first supply of continuous electrical current, opening new worlds of science.

GALVANI'S FROGS

In his famous experiments on dead **frogs' legs**, **Luigi Galvani** was able to show that exciting them with some sort of **electrical stimulation** causes them to twitch. In his experiments Galvani used **arches made up of more than one type of metal**.

ACADEMIC DISPUTE

Galvani and **Alessandro Volta** were in **dispute as to the source of the electrical stimulation** that made the frogs' legs twitch. Galvani thought it came from within the animal tissue, but Volta believed it came from outside, and that the frogs were acting as detectors. Volta wanted to make an inorganic model of the system, with no animal tissue, to prove his point.

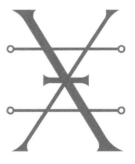

BIMETALLIC CONTACT POTENTIAL

Volta knew from earlier experiments he had repeated on himself that **placing two different metals on your own tongue creates a tingling sensation and a strange taste**. This is because the **saliva** on the tongue **acts as a conductive medium for electricity generated through the contact of dissimilar metals**, a phenomenon known as **contact potential**.

BUILDING A PILE

To create a model of the process, Volta stacked up **alternating disks of silver and zinc, placing between them cardboard soaked in salty water**. Each **silver-cardboard-zinc "cell"** generated only a **tiny current**, but by **stacking them up** in a pile, Volta was **able to generate a considerable current**. He had invented the **first battery**.

METAL HIERARCHY

Like any good engineer, Volta tried to improve his invention. Investigation showed that **different metal combos produced more current**, and he soon **switched to copper** in place of the **more expensive and less effective silver**.

EARLY ELECTRIC LIGHTS

Lighting technology changed little until the late eighteenth century, when engineers began to improve lighting, first with combustion technology and then with electrical technology.

LIGHT AND HEAT

Engineers seeking to improve lighting had two important factors to consider: the **temperature that the light-emitting material can reach** (higher temperatures usually mean more light); and the **efficiency with which fuel is turned into light**. **Electricity** would **revolutionize** both of these.

HOLLOW WICKS

In 1784 a classic example of engineering helped improve **pre-electric lights**, when the Swiss **Aimé Argand** patented an **oil lamp with a circular wick**, which **allowed air to reach the inside of the wick**. **Larger flames** led to **increased brightness**.

HEAT SPREADERS

Attempts to engineer **gas burners** in a similar way, with **spreaders that fanned out the burning gas**, ran into **problems** as the **metal in the spreaders conducted away heat and cooled the flames**. This was solved by use of **non-conductive soapstone** to make spreaders.

ELECTRIC ARCS

The **advent of the battery** provided a **supply of electricity**. English scientist **Humphry Davy** showed that **a spark could be made to bridge the gap between two carbon points**. This spark, known as an **arc**, gave off a **brighter light than had ever been seen before**.

BURNING CARBON

A **problem with the arc lamp** was that the **carbon points would burn away as it operated**, and **needed to be moved closer together**. The **first automatic electric control devices** ever created were **electromechanical regulators** that would **close the gap in response to changes in current or voltage**.

RESISTORS

A vital component of electrical circuits today, the resistor traces its origins back to the work of German scientist Georg Ohm in the 1820s.

MEETING RESISTANCE

Electrical experimenters of the eighteenth century noticed that **some materials** were **highly conductive**, while **others insulated one from electrical shock**. They further noticed that the flow of the electrical medium (what would today be called **current**) sometimes **met with resistance**.

WHAT IS RESISTANCE?

Resistance is the **tendency of a material to resist the passage of an electric current** and to **convert electrical energy into heat energy**.

OHM'S LAW

Ohm did his own experiments with **conductivity** in what were then called electrical "**chains**" (**circuits**), seeking **mathematical descriptions for the properties of conductive materials**. He was able to show that **resistance is proportional to the voltage and current in a circuit**; this is now known as Ohm's law, and **resistance is measured in ohms**.

RESISTANCE IS UTILE

Resistors are among the most common components of electrical circuits. They **reduce**, **limit**, and **split voltages**, **protecting components** and shaping the forms of electrical waves to **fit them for purpose**.

TYPES OF RESISTOR

The most common type of resistor is the **carbon resistor**, made from **mixing carbon particles with a ceramic mortar**. Other types include **wire coil and film resistors**. Resistors can be made to have a **fixed resistance value** or to be **variable**.

DYNAMOS

*Although the voltaic battery could supply a flow of current,
electricity remained a curio with few practical applications. The discovery
of the link between electricity and magnetism would change this.*

TWITCHY NEEDLES

In 1820 Danish scientist **Hans Christian Ørsted** showed that an **electric current flowing through a wire could deflect the magnetic needle of a compass**.

INDUCTION

In 1831 British scientist **Michael Faraday** showed that **moving a wire past a magnet could induce an electrical current in the wire**. Clearly electricity and magnetism were two sides of the same coin: **electromagnetism**.

GALVANOMETERS

One of the **first practical applications** of the discovery of **electromagnetic induction** was the creation of a **simple device for measuring the strength of an electric current by the degree to which it magnetically deflected a needle**: a **galvanometer**. The **ability to accurately measure electricity** would transform the field.

FARADAY'S DISK

Faraday soon constructed the **first electric generator**, in the form of a **copper disk that spun between the arms of a horseshoe-shaped magne**t. Other generators soon followed, but their inventors found that because the **opposing poles of the magnets generated currents in opposite directions**, they were **producing alternating current**.

COMMUTATORS

To **convert the alternating current to direct current**, French engineer **Hippolyte Pixii** invented the **commutator**, a device that **reverses one of the current directions** to produce a direct current. A generator with a commutator is called a **dynamo**.

ELECTROPLATING

Dynamos could now supply **useful current**; one of the **first industrial applications** was for **electroplating**, which used current to **deposit a thin layer of valuable metal such as silver or gold onto less valuable metals**.

TELEGRAPH

Electricity found its "killer app" in the shape of the telegraph, creating a new industry that spurred an explosion in electrical engineering.

SEMAPHORE

The **first telecommunication systems** used optical signals, such as the **coded-flag semaphore system** used by Napoleon, or the shutter-based system used by the **British Royal Navy**. But they had **major drawbacks**—they **could only be used at day and in good weather**.

CURRENT THINKING

The invention of the **battery** ensured a **supply of electric current** and Ørsted's **discovery that an electric current can deflect a magnetized needle** meant that the pieces were in place for a **signaling system using electricity**.

TWIN TRACK

In 1837 British inventors **William Cooke** and **Charles Wheatstone** patented an **electric five-needle alphabetic telegraph**. In 1839 it was incorporated into a twelve-mile stretch of the **Great Western Railway**. **Railways and the telegraph would henceforth be inseparably linked**.

MORSE CODE

Samuel Morse was a US artist and inventor who heard about electrical **developments in Europe** and dreamed up a **system for electric telegraphy**. He devised a **binary code** and with his partners invented the **Morse key** for **transmitting** it. In 1844 he sent the **first message by wire in the United States**: "What hath God wrought!"

UNDERSEA CABLES

Wires quickly stretched out around the world, including **transatlantic cables**. The first attempts at these failed due to a combination of poor cable design and misunderstanding of the science of electricity, but when Scottish scientist and electrical engineer **William Thompson** took over, the transatlantic cable was a success.

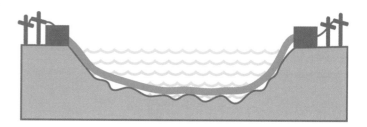

SIEMENS SELF-EXCITING DYNAMO

This key technology in generating high-power electricity from mechanical energy was pioneered by the German engineering colossus Werner Siemens (1816–92) in 1866.

UNDER POWERED

More than thirty years after **Faraday's discovery of electromagnetic induction**, **dynamos** (generators producing direct electrical current) remained **weak and underpowered**. To produce the **magnetic field** needed for **induction**, these dynamos used **permanent magnets**, which had to be **heavy** to produce even a **weak current** (a 4,400-pound magnet gave just 700 watts of power) and were **easily demagnetized** by the vibration of the dynamo.

SELF-POWERED

German engineer Werner Siemens, who had **already improved electromagnetic devices** for the **telegraph**, set his mind to improving the dynamo. He knew that **electromagnets** (magnets created by passing **electric current** through **coils of wire**) could be **very powerful**, and cleverly **devised a dynamo that created its own magnetic field from its own electrical output**.

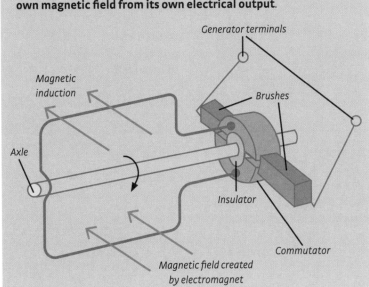

Generator terminals

Magnetic induction

Brushes

Axle

Insulator

Commutator

Magnetic field created by electromagnet

SELF-EXCITEMENT

But how could the **dynamo-electric effect** get started in the first place, if the device's electromagnet depended on its own electrical output, which in turn depended on its magnetism? Siemens showed that residual magnetism in the iron in the electromagnet was enough to get the ball rolling, through a kind of boot-strapping effect he called "self-excitement," or the "dynamo-electric principle."

ELECTROMAGNETS

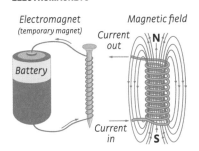

Electromagnet
(temporary magnet)

Current out

Magnetic field

N

Battery

Current in

S

AC vs DC

The war of the currents, between Thomas Edison, who promoted direct current (DC) and George Westinghouse, an advocate of alternating current (AC) was won by AC, and today alternating currents bring electricity to almost every building in the world.

- **Before 1800** static electricity was the only form of electricity that could be generated and studied

- **1799** Alessandro Volta invents the battery, which produces direct current (DC); i.e., it flows one way only

- **1805** Humphry Davy invents the arc lamp

- **1879** Joseph Wilson Swan demonstrates the first incandescent electric lightbulb

- **1883** US inventor Thomas Alva Edison forms a partnership, the Edison and Swan United Electric Light Company

- **1889** Edison powers the first electric chair with AC, trying to prove it is dangerous and could be used for executions

WAR OF THE CURRENTS

For **street lighting** in New York City, **Edison used direct current (DC), while George Westinghouse used alternating current (AC)**, in which the current flows one way and then the other in rapid fluctuation. The men having been in competition since 1886, the conflict between them turned into a **huge media war** by 1888, dubbed the **War of the Currents**.

TURBINE GENERATOR

Turbines produce alternating current. A transformer can **step the current up to a high voltage**, which is **much more efficient for long-distance transmission**.

MAJOR POWER LINES

The **voltage** is then **stepped down by another transformer at the other end**, but some claimed that the **high voltage was a danger to the public**.

AC RULES

Today almost all power is distributed by AC. The voltage for **domestic supplies varies** from country to country: in the **UK** it is **230 volts**; in the **US, 120**; in **Japan, 100**.

Voltage (V)

DC

Time (ms)

AC

THE LIGHTBULB

The landmark technology that marked the dawning of the Power Age, the lightbulb—more properly, the incandescent filament lamp—was a triumph of electrical engineering.

HARSH LIGHT

Arc lights were widely available by the 1870s, but their **light was too harsh for indoor and domestic use**. US inventor **Thomas Edison** was one of many determined to invent what he called a "**mild electric light**."

HOT WIRE

The phenomenon of **resistance** means that many materials **turn electrical energy into heat energy**. A **wire** of such material will **heat up until it glows**, and it was clear that this might form the **basis of a lighting system** if it were possible to **stop the material from burning**.

WIRE AND BULB

Invention of the lightbulb required two things: a **material for the wire that can repeatedly be heated until it is white hot**; and a **way to seal it inside a glass vessel** (a bulb) from which all the air has been expelled to create **a vacuum**. In the mid-1870s, the **mercury vacuum pump** meant the latter was available.

CARBON CANDIDATES

Among the many inventors who pursued the lightbulb, the best known were the Americans **Thomas Edison** and **Hiram Maxim** and the **Britons Joseph Swan** and **St George Lane Fox-Pitt**. They all identified carbon as the best material to use for filaments, but chose differing sources:

Edison:
Bamboo

Lane Fox-Pitt:
Grass

Swan:
Cotton

Maxim:
Paper

BETTER BULBS

Later bulbs used **metals with very high melting points**, such as **osmium**, **tantalum**, and, by 1911, **tungsten**. Adding **halogen gas** to the vacuum in the bulb **reduces evaporation of tungsten** and **lengthens the life of the filaments**.

THE TELEPHONE

The advent of the telegraph spurred many to consider how to send sound-reproducing signals over the wire, but it would be Alexander Graham Bell who would succeed.

HARMONIC TELEGRAPH

Alexander Graham Bell was a **professor of vocal physiology** whose expertise in the field of **sound waves** had led him to work on **electrical waves**, specifically, a way to **send multiple electrical signals along the same wire by pitching them at different frequencies**, to create a "**harmonic telegraph.**"

TELEPHONE CHECKLIST

A working telephone needed a means to convert **sound into electrical signals and back again**, and a way to **send these signals down a wire**, while **allowing other signals to travel in the opposite direction**.

MIXED SIGNALS

While engaged in this research, Bell clearly heard a **sound being transmitted over the telegraph** and realized that **his technique**, which **modulated both the frequency and amplitude of electrical waves** in the wire, could be used to **transmit sound and therefore speech**. It could also allow **duplex (two-way) transmission of signals**.

TELEPHONE BASICS

The standard telephone technology that emerged **converted speech to electrical signals** using a **diaphragm over a cup packed with carbon grains**. Sound waves **compressed the diaphragm**, which in turn **varied the density of carbon grain**, which in turn **varied the strength of a current flowing through the cup**. A **receiver** at the other end **reversed the process** using a **diaphragm vibrated by an electromagnet**, itself **controlled by the incoming signal**.

PATENT PALAVER

In 1876 Bell filed a patent for a telephone with the US patent office—just hours before a similar claim by inventor **Elisha Gray**. It has since been claimed that there was **funny business at the patent office**, and eventually **Bell would buy out Gray's company**.

THE PHONOGRAPH

Thomas Edison's favorite invention was the phonograph, a device for recording and playing back sound, which would create the recording industry.

LAMP-BLACK RECORDINGS

Since the early 1800s, many inventors had used the **vibrational nature of sound waves** as a way to **record them**. A popular medium was a **thin layer of carbon** (known as **lamp-black**) **deposited on paper** or another substrate, **typically wound around a cylinder**. A **needle vibrated by sound** would trace a pattern in the lamp-black as the cylinder revolved beneath it.

TELEGRAPH OFFSHOOT

Like many advances in electrical engineering at this time, the **invention of the phonograph was an offshoot of work on the telegraph**.

LEAVE A MESSAGE

In 1876 **Edison** was working on ways to **record telegraph messages**. One idea he pursued was to **convert the electrical pulses into traces scratched on paper**, **foil**, or **wax by a stylus**. He even recorded a **telephonic transmission** in this way, and was amazed to hear, when **drawing the paper back underneath the stylus**, a **faint reproduction of the original sound**.

EDISON'S PHONOGRAPH

Seizing the opportunity, Edison directed his team to develop the phonograph, producing a **working model** by 1877. It **scratched indentations on foil** (later changed to **wax**) **wrapped around a cylinder** and used a **trumpet to amplify the playback**. Edison **developed an entire recording industry around his invention**.

GRAMOPHONE

In 1888 German inventor **Emile Berliner** patented a similar device that **recorded sound on disks rather than cylinders**. These were **easier to reproduce and handle**. Because **Edison owned the rights to the name phonograph**, Berliner called his device the **gramophone**.

NIKOLA TESLA

Possibly the greatest electrical engineer of all time,
Tesla was a visionary but also an eccentric and a crank.

COMING TO AMERICA

Nikola Tesla was a Serb, **born** in 1856 in what is now **Croatia**, who eventually **emigrated to the United States** and **spent most of his life there**.

FAMOUS INVENTIONS

Tesla is celebrated for many **inventions and achievements**, including his design of the **polyphase induction motor**, the **hydroelectric plant at Niagara Falls**, the **Tesla coil**, the **bladeless turbine**, and **radio remote control**.

WILD SCHEMES

Tesla dreamed up many schemes and ideas that were never realized, or perhaps realizable. They included a **plan for a giant hoop around the Earth**, a **resonance device that could destroy distant targets**, a **particle beam death ray**, **wireless power transmission**, and **communication with other worlds**.

ECCENTRICITIES

Tesla was afflicted with **strange phobias and compulsions**. He was **terrified of dirt** and demanded **sterilized cutlery**; **pearl earrings** and other people's hair made him sick; he **performed actions in multiples of three**; and he even **fell in love with a pigeon**.

POLYPHASE INDUCTION MOTOR

*A beautifully elegant example of electrical engineering, the polyphase motor designed
by Nikola Tesla formed the basis for the architecture of the electrical industry.*

POLYPHASE EXPLAINED

A polyphase induction motor
uses **two or more streams of
AC current**, out of phase with
one another, to **drive a motor by
electromagnetic induction**.

BRUSH WITH DISASTER

Tesla got the idea for his polyphase motor after seeing a demonstration of a
Gramme dynamo. To **convert AC into DC**, the dynamo used **commutator
brushes** (metal brushes that maintained a connection without being fixed,
thus allowing the rotor to revolve freely). Brushes had many **drawbacks**,
including **sparking**, **dust**, **arcing**, **needing adjustment and replacement**,
and **general inefficiency**.

USING AC

The commutator was necessary
because **no one had figured out
how to use AC current to make
a motor go in one direction**.
Not only did the current **reverse
direction many times a second**,
but it **peaked and troughed** and so
**did not give a continuous power
supply**.

OUT OF PHASE

Tesla realized that supplying **multiple phases of current** would mean
that as one **phase was dying away**, **the other would be ramping up**,
giving a **continuous supply**.

CHASING ITS TAIL

His polyphase motor used
**electromagnets spaced at
intervals on a stator ring around
a revolving armature or rotor**.
The **phasing of the AC current**
means that the **magnetic field
they generate** is **always circling
around the stator ring, dragging
the rotor around after it**. This
generates the **rotary power of the
motor**.

ELECTRIC POWER GENERATION

Today the entire world runs on a grid of interconnected power stations,
which traces its origins back to a single New York block in 1882.

ULTIMATE POWER

Electric generators **convert some other form of energy into electricity**.
Since most generators use **turbines** to revolve wires in **magnetic fields**, they
require a source of **kinetic energy**, the nature of which varies. **Hydroelectric
power** uses energy from **running and falling water**; **steam turbines** get
the energy to heat water into steam from **burning fossil fuels** such as **coal** or
gas, or from **nuclear**, **geothermal**, or **solar thermal energy**. Other sources
include **wind** and **wave** power.

PRIVATE POWER

The first generators were for **private use**, usually in an **industrial setting**.
In 1879, for instance, at **Dolgeville** in **New York state**, a **mill** was fitted with
a **dynamo** to provide electricity to industry. Existing **hydropower** provided
obvious locations for electricity generation.

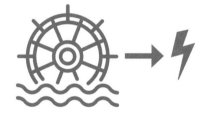

FIRST EDISONS

The **first generators for public
supply of electricity** were
installed by **Edison's company**
in 1882. The **first was in London**,
and proved **unsuccessful**, but
one installed on **Pearl Street in
Manhattan**, to **power lights in a
half-block radius of the financial
district**, proved to be the **model
for the future power grid**.

STEAM TO ELECTRICITY

Edison designed his
own **giant dynamo**,
nicknamed **Jumbo** after
a famous circus elephant,
and **installed a
number of these in his
power stations**, where
they were **driven by high-
speed, coal-powered
steam engines**.

RADIO

The invention of radio is a classic example of the role of electrical engineering in taking a scientific discovery and turning it into a practical technology.

MAXWELL

In 1864 Scottish scientist **James Clerk Maxwell** theoretically proved that an **oscillating electric current produces electromagnetic waves that travel at the speed of light**.

HERTZ

In 1886 German physicist **Heinrich Hertz** provided **experimental proof of these waves' existence**, using a **spark gap** between **two charged globes** to generate **a burst of electrical oscillation**, and a **loop of wire** on the other side of the room in which the **electromagnetic waves** would induce a **kind of echo**, like a wineglass that rings in response to a loud tone at the correct frequency. The current thus induced **generated another spark across another spark gap**.

LODGE

This **wireless transmission** of a signal had obvious potential applications. One of the first to explore this was British physicist **Oliver Lodge**, with a device that could **detect the radiating waves and convert them into an electrical pulse**. It consisted of a **tube of iron filings**, which would **stick together** or cohere **in response to an electromagnetic wave**: a **coherer**.

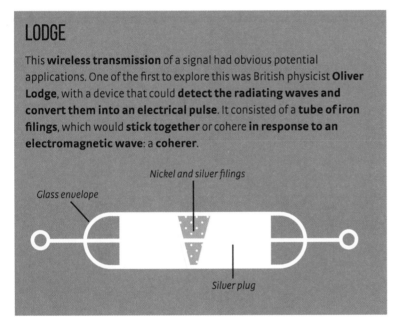

Glass envelope

Nickel and silver filings

Silver plug

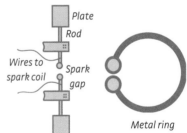

Plate

Rod

Wires to spark coil

Spark gap

Metal ring

MARCONI

Young Italian radio pioneer **Guglielmo Marconi** read about the discoveries of **Hertz** and **Lodge** and worked to **increase the distance over which signals could be sent**. He found ways to **boost the power of the transmitter** and the **sensitivity of the receiver**, and made equipment **simple enough for anyone to use**.

TRANSATLANTIC BROADCAST

In 1901 Marconi stunned the world by **sending a radio message from England to North America. Wireless telegraphy would soon become ubiquitous.**

DIODE VALVE

*A simple adaptation of the lightbulb would
revolutionize radio and usher in a new age of electronics.*

THE EDISON EFFECT

Researching his lightbulb, **Edison** accidentally stumbled on a phenomenon known as **thermionic emission**, which he promptly patented as the **Edison Effect**. It is where a **heated metal element gives off electrons**.

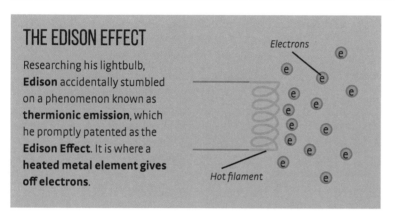

Electrons

Hot filament

ONE-WAY TRAFFIC

Edison found that if he **introduced another wire into the bulb**, electrons from the heated filament could cross over to it and cause a current to flow, but only if the second wire had a positive charge.

VALVE

British electrical engineer **John Fleming** realized in 1904 that this modified bulb constituted a sort of **electronic valve**, allowing **current to flow one way but not the other**. Thus it could **convert alternating current into direct current**. This property is known as **rectification**.

RADIO DETECTOR

Fleming was working with **Marconi** and realized that the valve, **known as a diode because it** had two electrodes in it, would make a **much clearer detector of radio waves than a coherer**.

AM RADIO

Diodes became even more useful with the advent of **sound broadcasting**, since sound information was **carried by the radio waves in the form of amplitude modulation**, which **required rectification before it could be converted back into sound**.

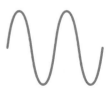

COMPUTER COMPONENT

Valves would later prove **essential to computing**, since they can be used to make **logic gates**.

TRIODE VALVE

The diode was useful but underpowered. Electrical engineer
Lee de Forest transformed it with a simple bit of engineering.

WEAK SIGNALS

The **current that came out of a diode could be only as powerful as the current that went in**, and **radio signals** tended to be **very weak and low energy by the time they reached a receiver**.

POWER BOOST

In 1906 de Forest added a **third electrode** to the diode valve **to create a triode**. It consisted of a **wire grid**. The **holes allowed the electrons to pass through**, but by **varying the potential of the grid** it was possible to **greatly increase the flow of electrons from the emitter to the anode**. Thus, the triode could **amplify signals**.

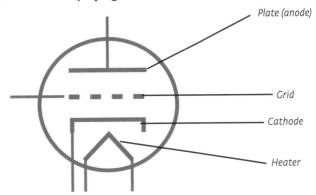

Plate (anode)

Grid

Cathode

Heater

AUDION

De Forest patented his triode, calling it the **Audion**. But the Audion had a **problem**: although like all valves it was a **vacuum tube device, the vacuum inside was "soft"—traces of gas remained**. This made the Audion **weak** and **unreliable**.

HARD VALVES

In 1913 the **first hard valve** was created, with a **hard vacuum in the tube**. These became **standard in radios around the world**. **More electrodes were added** to valves to create **tetrodes** and even **pentodes**.

VALVE COMPUTERS

Valves reached the **high point of development** in the 1940s when they were **used in huge numbers** to create the **first computers**, such as **ENIAC**. They would soon be **superseded**, however, by the **transistor**.

TELEVISION

A system for recording, transmitting, and reproducing moving images, television is credited to many inventors but, in reality, it was the product of collective engineering.

NEEDFUL THINGS

Television technology depended on **four elements**: a **way to change light into electric current**, and a **way to change it back**; a **device to scan an image into smaller elements**; and a **way to amplify weak signals to be strong enough to use.**

PHOTOCELLS AND SPINNING DISKS

In 1873 it was discovered that the **metal selenium** produced an **electrical current when illuminated**. In 1884 German inventor **Paul Nipkow** invented a **spinning disk with a spiral of holes**, which **mechanically scanned an image into lines**. Turning electric signals back into light required a **strong electric light source**: the **neon light** invented in 1903 was the favored choice for early inventors. **De Forest's Audion triode** provided the final piece of the puzzle.

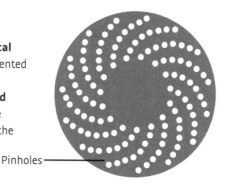

Pinholes

MECHANICAL TELEVISION

Scottish inventor **John Logie Baird** devised a **working demonstration of mechanical TV**, so called because the **image was scanned by a spinning disc** (i.e., by a machine with moving parts). The **holes** in the disc **admitted light from a target**, which **fell on a photoreceptor** and were **converted into electrical signals** and **sent by radio wave to a receiver**. Here the **brightness of a lamp varied in response to the incoming signals**, shining through a **revolving disc synchronized with the camera disc.**

ELECTRONIC TELEVISION

First developed by the American **Philo Farnsworth**, electronic television depended on the **cathode ray tube**, in which a **beam of electrons** (the cathode ray) **"paints" an image onto a phosphorescent screen.**

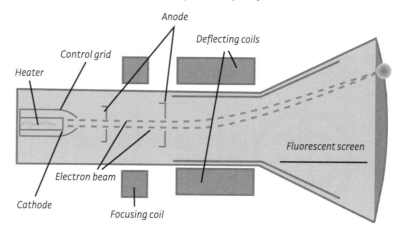

Anode

Deflecting coils

Control grid

Heater

Fluorescent screen

Electron beam

Cathode

Focusing coil

RADAR

War stimulates the application of engineering talent to turn science into technology; radar is a classic example.

RA.D.A.R.

Radar was originally an **acronym**, standing for **RAdio Detection And Ranging**.

RADIO INTERFERENCE

Radio users had long known that **passing vehicles** could **interfere with radio reception**, and this suggested obvious **detection applications**. The **first radio echo device** was patented in 1904. But the **technology was not up to the task**—in particular, the **radio wavelengths** were **too long to be useful**.

ECHOLOCATION

Radar is similar in principle to **echolocation**. **Radio waves** are **emitted, bounce off a target**, and the **echo is detected**. Because the **speed of the radio waves** is **constant**, it is **possible to work out from the delay the distance to the target**.

MAGNETRONS

By the 1930s, **shorter wavelengths** could be generated, although in 1937 the **British system was still using waves over 3.3 feet long**, which needed **large antennae** that **restricted deployment**.

The invention of the **cavity magnetron** by the British in 1940 made available **waves less than an inch long**, greatly **improving sensitivity and accuracy**.

BASIC RADAR

The basics of a radar system include a **magnetron** to generate **short wavelength signals** (actually **microwaves** rather than radio waves); an antenna to transmit these, usually as a beam or plane; an **antenna to pick up echoes**—usually the same antenna, with a **duplexer** fitted that switches the antenna between transmitter and receiver; a **processing unit to extract information from the echo**; and a **display device to allow an operator to visualize the info**.

1 = Magnetron
2 = Duplexer
3 = Antenna (broadcasting)
4 = Antenna (receiving)
5 = Processor
6 = Display

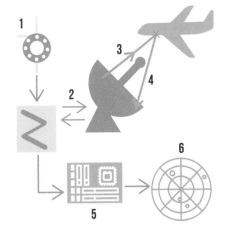

MICROWAVE OVEN

A classic example of an accidental discovery, microwave cooking was discovered by US engineer Percy Spencer in 1946.

THE MELTED SNACK

Spencer had worked on **radar** during the war and was **trying to improve the power output of the magnetron**, a device that **generated microwaves using resonant cavities** (a bit like a whistle but with EM waves instead of sound). After one test he noticed that the **snack bar in his pocket had melted**.

NUTS TO CHOCOLATE

Many versions of this popular legend say the snack was chocolate, but in fact it was a **peanut bar**, a detail that was important because the **nut bar had a much higher melting point than chocolate**. Spencer **tested an egg**, which promptly **blew up in his face**.

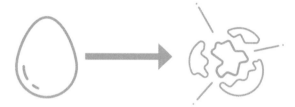

MICROWAVE COOKING

A microwave oven works by **passing high-frequency, short wavelength EM radiation through solids or liquids. Water molecules**, in particular, **absorb microwaves**, so they **heat up quickly, cooking the rest of the food**.

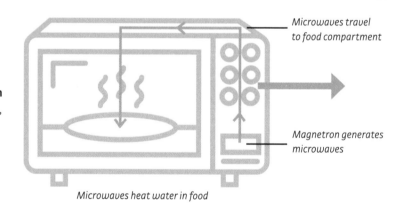

Microwaves travel to food compartment

Magnetron generates microwaves

Microwaves heat water in food

DOOR GRATE

A microwave oven has **five grounded metal sides**, so that microwaves that are absorbed pass on their energy as electric charge that dissipates. It is useful to be able to see through the sixth side (the **door**), so it is fitted with a **grounded metal mesh with holes smaller than the wavelength of the microwaves** (typically c.5 in). **Visible light** has **wavelengths shorter than the diameter of the holes**, and so can **pass through**.

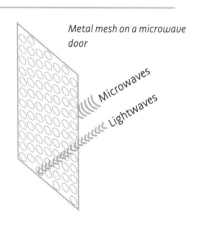

Metal mesh on a microwave door

Microwaves

Lightwaves

TRANSISTOR

The transistor was the essential component allowing the birth of microelectronics and the world of modern IT.

VALVE REPLACEMENT

Thermionic valves had proven the principles of **rectification** and **amplification** in electronics, but they were **bulky**, **power-hungry**, and **unreliable**. **Semiconductors** seemed to offer a better alternative, if they could be made to work.

SEMICONDUCTORS

Semiconductors are **materials that can switch between being conductors and insulators depending on external factors**, such as **applied voltage**.

BELL LABS

Bell Labs, of the US telecoms giant **AT&T**, was the **premier research institute of the early electronics age**. It was Bell that had **perfected de Forest's triode amplifier**, and it was Bell that would now realize the transistor.

TRANS-RESISTOR

The name "transistor" was coined at Bell as a sort of contraction of "**trans-resistor**." Essentially a transistor is a type of **multielectrode valve that does not need a vacuum tube, heating element, or any of the other things that make thermionic valves unreliable.**

POINT CONTACT TRANSISTOR

In 1947 at Bell Labs, scientist **John Bardeen** and engineer **Walter Brattain** devised the **first transistor**, with **two gold point contacts** very close together, **in contact with a layer of semiconducting germanium**.

SANDWICH TRANSISTOR

Bardeen and Brattain's boss, **Bill Shockley**, promptly invented a **simpler and easier-to-make transistor**, with a **"meat" layer of electron-poor semiconductor between two "bread" layers of electron-rich semiconductor**. Applying a **positive voltage** to the meat **causes an amplified, one-way flow of current** between the bread layers.

INTEGRATED CIRCUIT

Devised independently by two men at the same time, the integrated circuit or chip is a way to make all of the components of a circuit at the same time, out of the same material, meaning it can be made much smaller.

SIZE MATTERS

Transistors had rapidly replaced thermionic valves and become **ubiquitous in electronics**, with devices such as the **transistor radio** having far-reaching effects on, for instance, **youth culture and music**. But transistors were **reaching a size limit**, beyond which **connections to other circuit components could no longer be made by hand**.

CRYSTAL MATH

In July 1958 **Jack Kilby**, an engineer at **Texas Instruments**, realized that **if all the components of a circuit**, such as **capacitors** and **resistors**, could be **made from the same semiconductor material as the transistors** then they **could all be machined from the same piece of material**. Just a few months later, **Robert Noyce** at **Fairchild Semiconductor** independently came up with a **similar but faster technique using photo-etching**.

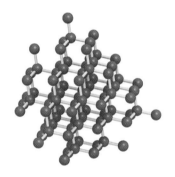

STENCILS AND MASKS

Modern microchips are made with an **advanced version of Fairchild's photo-etching "planar process."** A **single large wafer of silicon is used to make hundreds of chips**. **Layers** of **conducting, insulating,** and **semiconducting materials** are **laid** down **on top of one another** by using **masks** laid down with **stencils**, which then act as **stencils for other layers before being dissolved**.

LAYER BY LAYER

The **base layer** is **electron-rich P-type silicon**. Insulating **silicon dioxide** is laid down **on top of this**, then a **stencil is used to apply a mask**, which **protects some areas of silicon dioxide** but **leaves others to be dissolved**. Other layers, including **conductive polysilicon** and **electron-poor N-type silicon**, are **added**. The final additions are **tiny strips of metal to connect components**.

LASER

The laser, originally an acronym for Light Amplification by Stimulated Emission of Radiation, is the hi-tech product of a marriage between advanced engineering and quantum physics.

IN STEP

Quantum physics predicts that it **should be possible to produce electromagnetic radiation that is coherent**, which means **all of the same wavelength** and **perfectly in phase**. One way to achieve this is to **pump energy into atoms** in such a way as **to make them emit photons**, and then to marshal these into a **coherent beam**.

MASER

In 1953 **Charles Townes** and his team managed this feat with **microwaves**, with a technique they called **Microwave Amplification by Stimulated Emission of Radiation**, or **Maser** for short.

LASER COMPONENTS

To make a laser you need a material, known as a **lasing medium**, which can be made to **pump out photons of a single wavelength**; a **power source**; a way **to get them all to march in step in the same direction**; and a **device to focus the beam**. Lasers are also often equipped with **collimators**, which **focus the beam** to a **narrow point of extreme intensity**.

FIRST LIGHT

The **first laser** was created in 1960, by **Theodore Maiman** at the **Hughes Research Lab**, in Malibu, California. He used a **rod of ruby** as the **lasing medium**, wrapped in a **coil of xenon flash lamp** (essentially a sophisticated fluorescent light tube). The **lamp pumped energy into the rod**, which was **flat and mirrored at each end**, so that the **photons bounced back and forth along its length, cohering into a beam**.

LASER APPLICATIONS

Apart from **military interest** in **laser weapons**, lasers have myriad other applications. In **surgery**, lasers are used to make **ultrafine incisions** of just 0.5 microns wide (a human hair is ~80 microns wide). Other uses include reading **optical disks**, **microscopy**, **barcode scanning**, and **fiber optics**.

MOON MEASURES

Lasers are used in **surveying**, including measuring the **distance between the Earth and Moon. Laser reflectors** left **on the Moon** by Apollo astronauts have **made it possible to measure this distance to within six inches**.

HUMAN–COMPUTER INTERACTION

The field of engineering IT to facilitate and enhance its operation by users is known as human–computer interaction (HCI). Its greatest exponent was Doug Engelbart.

PUNCH CARDS

The **original computers** were **operated by punch cards**, a bit like old-time player pianos. Later computers had **keyboard inputs** but required, at a minimum, the ability to negotiate **complex and nonintuitive programming languages**.

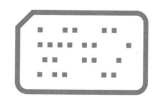

BOOTSTRAPPING

US engineer **Douglas Engelbart** believed that **human productivity and computer ability would advance in synergy** when **users and their tools were properly aligned**. During the 1960s he used a **systems engineering** term, "**bootstrapping**," to describe how **improving the tools to implement technology would advance the technology**, feeding back to **more improvement in the tools**, and so on.

NLS

At his **Augmentation Research Center** in California, Engelbart and his team devised a **computer hardware and software system** called the **oN-Line System** or **NLS**. It pioneered features now ubiquitous in IT, such as the **computer mouse**, **graphical user interfaces** with **windows**, **collaborative file sharing**, and **hypertext links**, among others.

THE MOUSE

The computer mouse had its **origins in trackballs** developed in the **Second World War** to **facilitate human–computer interaction**. In 1963 Engelbart sketched out an idea for what he initially termed a "**bug**"; a **prototype was built in 1964**. The **cord linking it to the computer looked like a tail**, hence the name "mouse."

THE MOTHER OF ALL DEMOS

In 1968 in San Francisco, Engelbart gave a **demonstration of his NLS system**, which has since become known as "**the mother of all demos**." It **helped disseminate the HCI principles** that would **transform the personal computer** and lead, for instance, to the **creation of the Apple Macintosh system**.

PERSONAL COMPUTERS

*The development of cheap programmable microchips in the 1970s
stimulated production of the first personal computers.*

MOORE'S LAW

Gordon Moore was an engineer at **Fairchild**, an early leader in
the production of **integrated chips** (ICs). In 1964 he noted that the
number of transistors that were **being fitted onto each chip
was doubling each year**, an observation that became **codified
as a projection**, known as "**Moore's Law.**" The introduction in the
late 1960s of **metal-oxide semiconductors in chips** ensured that
the law would hold—more or less—for years to come, dramatically
boosting the capabilities of single microchips.

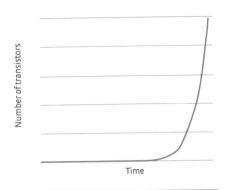

COMPUTER ON A CHIP

In 1968 **Moore** and **Noyce** set up **Intel Corporation** to make chips for
themselves, and in 1969 Intel engineer **Marcian Hoff** designed a **general-
purpose circuit** that could be programmed, to operate alongside **software
stored in memory chips**. By the mid-1970s the "**computer on a chip**" was
available for as little as $100.

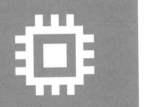

THE ALTAIR

Almost all computers were made for **business** and/or **large institutions**. But
there was also a thriving community of **hobbyists**, who put together their
own **components** to **make personalized systems for fun and research**. In
1974 a group of hobbyists created the **Altair**, a **personal computer kit**. **Bill
Gates** and **Paul Allen** wrote **software** for it, and went on to found **Microsoft**.

APPLE

Steve Wozniak and **Steve Jobs** founded **Apple** and marketed their **first
personal computer** (PC) in 1976. Jobs was especially excited by **Doug
Engelbart's ideas** and by 1983 Apple were producing computers with
keyboards, **integrated monitors**, **floppy disk drives**, **mice**, and **graphical
window interfaces** with **drop-down menus**. Meanwhile **IBM** had got into
the PC market in 1981, **commissioning Microsoft to write their operating
system**, **MS-DOS** (Microsoft disk operating system).

SEARCH ENGINES

The digital universe has added new aspects to the concept of engineering, as with the virtual machines created to explore the internet and the Web.

THE INTERNET

The roots of the internet lie in the work of US engineer **Claude Shannon**, the **"father of information theory,"** and network theory. This led to the development of **packet switching**, and in 1969 to the **ARPAnet, progenitor of the internet**.

WORLD WIDE WEB

An **information space or distributed information system** based around but **not synonymous with the internet**, the World Wide Web (WWW) was created around 1990 by British scientist **Tim Berners-Lee**. He put together a series of key elements: a **standard convention for network addresses**, so that **hyperlinks** could connect different **information sources**, **text**, and **media**; a **simple mark-up language (HTML)** to **format the information**; and a **transfer protocol** for the **transmission of information between servers and client users**. He also developed the **first Web browser**, a **software tool** to allow the user to use the WWW.

COMPONENTS OF A SEARCH ENGINE

Pseudo–search engines scan source material line by line to **check for precise matches** to a search query. **True search engines** are much **more sophisticated**, and consist of at least three different stages: **data prep**, **fulltext indexing**, and **search**. Data prep is carried out by **"spiders"** or **indexers**. Fulltext indexing makes a **record of all the content on the indexed pages**. Search is carried out by the **search engine proper**, which **generates results** that are then **presented to the user**. Complex **algorithms** are used to **rank search results** to best meet the search terms.

TOO MUCH CONTENT

At first, Berners-Lee and others **manually updated lists and directories of Web content**, but soon the Web became too big for this and **search facilities were developed**, beginning with programs like **Archie** in 1990. These soon developed into **search engines**, **virtual machines** for **cataloging**, **indexing**, **searching**, and **retrieving information** from Web pages.

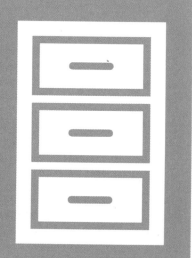

ARTIFICIAL INTELLIGENCE

The quest to engineer machine or artificial intelligence (AI) could be the greatest ever feat of human engineering, but it could also be the last.

TURING TESTS

Early advances in computer science, which led to **expert systems** such as **chess computers** that could play to very high levels, prompted many bold claims about the prospect of creating **machine intelligence comparable to human intelligence**. English computer pioneer **Alan Turing** proposed a different approach, suggesting that a **machine that appears to be intelligent** (able to pass a so-called "Turing test") **must be considered intelligent**.

ASPECTS OF AI

AI is not monolithic. The field includes areas such as **computer vision, natural language processing, pattern recognition, machine learning,** and **expert systems**.

WAYS TO BUILD AN AI

AI engineers have explored both **top-down** and **bottom-up** routes. Top-down approaches attempt to **encode thinking into symbolic languages that computers can deal in**. Bottom-up approaches involve **connectionism**, the **principle that intelligence may emerge from having enough of the right sort of connections**, such as in **massively parallel processing**.

NEURAL NETWORKS

Engineers build **units similar to single nerve cells**, allow them to **develop their own internal connective architecture**, and **"train" them by providing them with raw material and target outcomes**. Engineers may not know or understand what is happening inside the network.

THE SINGULARITY

Some AI engineers believe that **once a true machine intelligence is created**, the machine **will start to improve itself iteratively and exponentially**, almost immediately **progressing far beyond human abilities**. This will lead to an **inflection point in human history** known as the **Singularity**, which will lead to a **technological utopia** and/or the **end of humanity**.

QUANTUM COMPUTING

Quantum engineering holds the promise of a new type of computer with extraordinary capabilities, if the tricky macro-engineering can be achieved.

SUPERPOSITIONS

Quantum physics tells us that **particles can exist in more than one state simultaneously**—a superposition—until **forced into a single state by being observed or coming into contact with the outside world**, a phenomenon known as **decoherence**.

QUBITS

Classical digital computing is based on **bits, which can only exist in either of two positions or states**: 1 or 0, **on** or **off**. But a quantum computer uses **qubits**, which **can exist in a superposition of states**. A **single qubit can process multiple calculations simultaneously**.

ENTANGLEMENT

In a classical digital computer, **adding bits increases processing power arithmetically**, so a 32-bit computer is twice as powerful as a 16-bit one. Qubits, however, can be connected through a strange **quantum phenomenon called entanglement**, which links qubits in a kind of daisy chain that **multiplies their processing power exponentially**.

ENGINEERING CHALLENGES

The problem for quantum computing is that it is very **hard to keep qubits from decohering; particles** used for qubits **need to be kept at supercooled temperatures** or **confined in magnetic fields**. Building a computer around this is **technically extremely difficult**.

SPECIAL APPLICATIONS

Nonetheless, **some forms of quantum computer already exist**, and can be **used for special classes of problem with which classical computers struggle**, such as **modeling the behavior of atoms and molecules in drug development**, or **calculating optimal paths in optimization problems**.

ANCIENT MECHANICS

Working in a sphere related to but separate from the ancient Greek tradition of philosophy, mechanics were proud of their practical achievements and ingenious contraptions.

CTESIBIUS

A **third-century BC engineer and inventor** from Alexandria, Ctesibius was the **earliest named Hellenic master mechanic**. He made advances in **hydraulics** and **pneumatics**, the **study of water and air** (and, by extension, all fluids) **under pressure**. He was best known for his **accurate clock** and **hydraulic organ**.

KLEPSYDRA

As well as **sundials**, the ancient Greeks used **water clocks**, or klepsydra, where the **passage of time** is **marked by the level of water in vessels** as it **flows in or out. As the water level drops**, however, **so does the pressure** and **thus the rate of flow**, making such clocks **inaccurate**. Ctesibius is said to have devised a **device to maintain even pressure** and **thus keep the klepsydra accurate**.

AUTOMATA

Many ancient mechanics devised ingenious contraptions that **seem more like toys or curios than practical tools**. These included **sophisticated automata**. In fact, such devices **had a serious point**, as they were often used as the **basis for learned discourse**.

HYDRAULIS

Another pneumatic device was the **water organ** or **hydraulis**. Like modern organs, it **produced notes by blowing air through pipes** of different sizes. The **air supply** was mechanically supplied by a **pump**, and **pressure** was **controlled by a tank of water**. The **Romans** used them in the circus, where they held their entertainments.

WAR MACHINES

A primary concern of ancient mechanics was the construction of **ballistae**, **catapults**, and **siege engines**.

ANTIKYTHERA MECHANISM

When divers retrieved an unimpressive mass of corroded metal from an undersea wreck, they could never have guessed that they had found the most extraordinary artifact in ancient history.

MECHANICAL

THE WRECK

In 1900 sponge divers discovered the **wreck** of a first-century BC **Roman cargo ship** in waters off the island of Antikythera, near the northwest tip of Crete. As well as valuable **statues**, **coins**, and **other goods**, they recovered a lump of rock and barnacles, from which projected a **small piece of bronze**, which they labeled **Item 15087**.

ANCIENT CLOCKWORK

Eventually Item 15087 was identified as the remains of a **wooden-framed device stuffed full of sophisticated cogs and gears—ancient clockwork**. It was assumed to be some sort of **astrolabe**, a kind of ruler **to help work out the positions of heavenly bodies**, but it was not believed it could be any more complex, because the **first known mechanical astronomical calculator dated to a thousand years later**.

ANCIENT ORRERY

It is now known that the **Antikythera Mechanism** was a **highly sophisticated astronomical calculator** or **orrery**, a bit like an ancient computer. It could be **set by turning dials on the faces** and **cranking a handle that powered thirty-seven gears**, which **drove at least seven dials** showing the **movements of celestial bodies** and the **dates of eclipses**, among other functions. The **date on the machine** was set to the year 80 BC, **suggesting a date for the wreck**.

MECHANICAL MASTERS

It is likely that the Mechanism came from the island of **Rhodes**, which in the early first century was the **home of leading instrument makers and astronomers**.

THE AEOLIPILE

One of the great mysteries of the history of engineering was why the ancients failed to develop steam technology when we know they possessed it.

HERON OF ALEXANDRIA

Known to the Romans as **Hero**, he was a l**eading mechanic philosopher** of the Hellenic city of **Alexandria** in the first century AD. Heron developed the ideas of **Ctesibius** and **Archimedes** and wrote books on **automata**, **measuring**, **optics**, and **transporting heavy weights**.

STEAM DRIVEN

Perhaps his **most famous machine** was a curio with no known practical application: a **primitive steam turbine called the aeolipile**. In its **most basic form**, as described by the Roman engineer **Vitruvius** (first century AD), it consisted of a **kind of double-spouted kettle mounted on an axle**. The spouts were positioned so that when the spherical or cylindrical body of the aeolipile was heated, **steam would vent from the spouts, propelling the body in the opposite direction** by reaction (like a rocket). **Heron's aeolipile** was a little **more complex**, with **tubes connecting the revolving body to a boiler**.

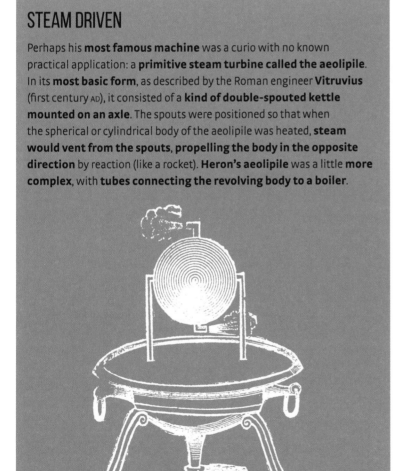

ALTERNATIVE HISTORY

Heron's aeolipile shows that the ancients **knew how to produce rotary power with a steam engine**. **Why did they not apply this technology to industrial or transport ends**, like their European descendants, and trigger an **ancient industrial revolution**? Explanations usually suggest that their **social economy was not structured to allow this**, particularly when they could make **slaves do hard labor for free**.

IMAGINARY APPLICATIONS

Heron's other achievements included the **improved water pump** and the **hydraulic organ**, which incorporated waterwheel-style rotary power generation. If he had thought to connect his steam engine to the former, he could have **created a powerful pumping engine** that would surely have found many applications in Roman hydraulic engineering, while **combining steam and rotary-wheel generators** could have **produced steam turbines**.

ZHANG HENG

Polymath Zhang Heng was the da Vinci of second-century AD China, whose engineering contributions included his famous earthquake-monitoring device.

RENAISSANCE MAN

Zhang Heng (AD 78–139), sometimes Anglicized as **Chang Heng**, had a long career in public service and was a celebrated **poet** and **historian**. He was also an important **mathematician**, **natural philosopher**, and **astronomer**.

PRECISION TIMEPIECE

Like some of his Hellenic counterparts, **Zhang Heng improved the accuracy and reliability of water clocks** by **regulating water supply**.

ARMILLARY SPHERE

He combined several innovations to create a **water-powered armillary sphere**, a complex mechanical model of the celestial sphere, which represents the motions of celestial bodies. One of his improvements led to the development of the **escapement**, a major component of **clockwork**.

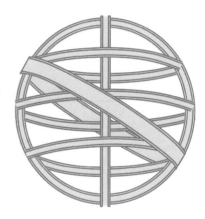

THE EARTHQUAKE INDICATOR

Zhang Heng's best-known invention was the ***Houfeng didong yi*** or **"earthquake weathervane,"** an early but highly sophisticated **seismoscope**, or earthquake-sensing instrument. It was **able to detect distant quakes and indicate the direction of the epicenter**.

MOUTH OF THE DRAGON

The seismoscope took the form of a **bronze urn**, around the exterior of which was a **ring of dragons indicating various points of the compass**. Detection of an earthquake would **trigger the release of a ball from the mouth of one of the dragons**, to **fall into the mouth of a bronze toad**.

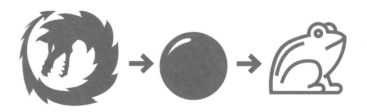

PUMPS

This unheralded technology has been surprisingly important to the development of science and industry.

ANCIENT PUMPS

Ancient engineers improvised the **first pumps** using **practical experience** rather than theory. The **shaduf**, a **counterweighted lever for raising buckets of water**, was **one of the earliest methods for moving fluid**. The **Archimedes screw** probably predates the Greek engineer by millennia; it **operates on the principle of the inclined plane**.

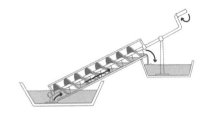

MECHANICAL

FORCE PUMPS

A force pump is one that **moves fluid using pressures in excess of atmospheric pressure**. **Ctesibius** of Alexandria designed one of the first known force pumps, using a **piston**.

PUMPS AND THE SCIENTIFIC REVOLUTION

Improvements in pump engineering, especially the introduction of **leather washers** to **reduce leakages**, meant that **soft vacuums** could be created. These were used for **scientific investigations** by pivotal figures such as Anglo-Irish chemist **Robert Boyle**.

TUBE TECHNOLOGIES

Better pumps that could achieve hard vacuums were **essential in enabling the development of tube technologies** such as **cathode ray tubes** and **X-rays**.

AIR PRESSURE

By medieval times, pumps were **essential in mining**, as miners delved ever deeper in search of ores but were limited by their ability to **pump water out of deep shafts**. But engineers were frustrated by the limit of thirty feet, the **maximum depth from which a mining pump would draw water**. In 1644 Italian scientist **Evangelista Torricelli** showed that this limit was imposed by the **equivalence between the weight of the atmosphere and that of a thirty-foot column of water**.

TYPES OF PUMP

The main two types of pump are **centrifugal pumps**, which **accelerate fluid**, usually with a **rotary impeller**, to create **pumping force**; and **displacement pumps**, which **draw fluid** in and **then expel it**, as with **piston pumps**.

CLOCKS

Timepieces were needed for astronomy and navigation, but regulating them became one of the great engineering challenges of the postmedieval world.

MECHANICAL ASTROLABES

The **great clocks of medieval times** were mechanical marvels, but they actually **served astronomical rather than timekeeping ends**, as they **were essentially clockwork astrolabes** (devices for calculating the motion of celestial bodies).

DRIVING FORCE

The **primary impetus** for clockwork, until the introduction of **springs**, was gravity. A **descending weight converted potential to kinetic energy**, which was **transmitted** to the motion of a pointer across a dial by a **series of cogwheels and gears**, a.k.a. a clockwork mechanism.

ESCAPEMENTS

A bar that rocked back and forth, alternately engaging the teeth of a wheel, known as an **escapement**, was an ingenious way to **restrain the force of the weight**, but until **Galileo** demonstrated the principle of the **pendulum**, there was no natural regulation to keep the escapement regular.

BETTER REGULATION

Dutch inventor **Christiaan Huygens** devised the **pendulum-regulated clock**, first built in 1657. There followed **centuries of attempts** to **improve the regulation of the motion of the escapement** and **compensate for variables** such as **thermal expansion**.

LONGITUDE COMPETITION

Navigators wanted a way to keep **accurate time independently of location**, to help work out **longitude**. In 1714 the British government offered **a prize to anyone who could devise an accurate clock**. After twenty-five years, clockmaker **John Harrison** perfected a highly accurate **chronometer** and claimed the prize.

STEAM TURBINES

The engines of Newcomen and Watt had proven that steam had enormous potential to drive useful work, but it remained a puzzle how to make the best use of this potent force.

MULTIPLE STAGES

One challenge was how to **extract more energy from the steam**. Making water change to steam takes **great amounts of heat**, but the **simple condensing engines** of Newcomen and even Watt **wasted a great deal of this energy**. One solution had been to have **more than one expansion vessel** so that the steam could be **made to do work multiple times**.

IMPULSE VS REACTION

Steam can **impart energy to a rotor** in two ways. In an **impulse engine**, the **steam pushes the rotor blades around**. In a **reaction engine**, the steam deflects off the blade in one direction, causing an **equal and opposite reaction in the other direction**. This is how a **rotating garden sprinkler** turns.

PARSONS

British engineer **Charles Parsons** (1854–1931) was the **first to create a successful steam turbine**, with a **reaction engine** that used **multiple sets of blades** to extract all the energy from the steam. His turbine powered an **electric generator**, and later **steamships**.

ROTARY MOTION

Another problem was that the **reciprocal motion of beam engines wasted a lot of energy**. A **turbine** that used **rotary motion**, as with **waterwheels**, would be better, but **steam has very different properties from water. Making a successful steam turbine** would prove to be **one of engineering's great challenges**.

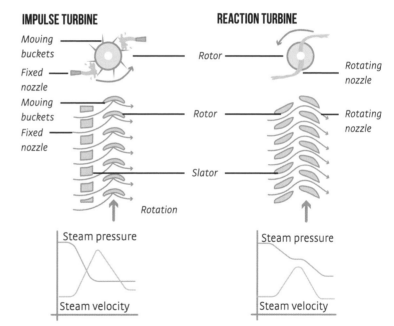

IMPULSE TURBINE

Moving buckets
Fixed nozzle
Moving buckets
Fixed nozzle
Rotation
Steam pressure
Steam velocity

REACTION TURBINE

Rotor
Rotating nozzle
Rotor
Rotating nozzle
Slator
Steam pressure
Steam velocity

MECHANICAL

CALCULATING MACHINES

Advances in science led to increasingly burdensome calculations, and the labor these entailed, and unreliability of the results, hampered the work of scientists. Could engineers come to the rescue?

DA VINCI'S MACHINE

Sketches from two **da Vinci** manuscripts suggest that he **conceived a calculating machine of some sort**, although skeptics say it was a **ratio machine** (a thought experiment to demonstrate the principle of ratios) that **could never have functioned in real life** because of **excessive friction**.

NAPIER'S BONES

The **earliest known mechanical calculating device**, **Wilhelm Schickard's calculating clock**, was based on **Napier's bones**, a **set of engraved rods** that could be **used as a calculation aid**, devised by Scottish mathematician **John Napier** in the early 1600s.

THE PASCALINE

In 1642 **Blaise Pascal** was an eighteen-year-old French math prodigy. To help his father, a tax collector, he devised a **mechanical addition calculator**. On the front of a rectangular brass box were **dials that moved internal cogwheels**. A **full rotation of one cog moved the cog to the left one-tenth of a revolution**, and so on. Called the Pascaline, the design was put into very **limited production**, but of around fifty units built **fewer than fifteen were sold**.

THE STEPPED RECKONER

German polymath **Gottfried Leibniz** learned the principle of the Pascaline and determined to build a **universal arithmetic machine** that could **multiply and divide as well as add and subtract**. His **Stepped Reckoner** used a special **cylindrical toothed gear** called the **Stepped Drum** or **Leibniz Wheel**. **Dials set the digits** and the **operations** were **performed by turning a crank**.

KEY CALCULATORS

Pascal's and Leibniz's designs formed the basis for most calculating machines for around 200 years. In 1884 US inventor **Dorr Felt** devised the **first successful key-input calculator**, although the **now-standard three-rowed ten-key format** dates to **Oscar Sundstrand's 1914 calculator**.

BABBAGE'S ENGINES

*The closest thing to a mechanical computer, the Analytical Engine,
was devised in the nineteenth century by British mathematician and engineer
Charles Babbage. A century ahead of its time, it was never constructed.*

COMPUTING BY HAND

Before the advent of machine **calculators and computers**, **calculators and computers were people** who worked out **lengthy and laborious sums by hand**. Babbage was infuriated by the inevitable **human errors** and in 1823 determined to devise a machine to do the calculations.

DIFFERENCE ENGINE NO.1

His first effort was a **clockwork machine**, or engine, that would **perform complex calculations** using a **mathematical technique** known as the **method of finite differences**. **Numbers** would be **represented by the rotation of cogwheels**, so that **cranking the engine would turn an input into an output**. Over ten years, Babbage spent £17,000 before he **ran out of money**. The engine was eight feet high and had 25,000 parts.

DIFFERENCE ENGINE NO.2

In 1847 Babbage drew up plans for a **simplified Difference Engine**. He failed to get government funding and the **project was never realized**.

THE ANALYTICAL ENGINE

In 1834 Babbage began work on plans for a **more ambitious device**, an **analytical engine** that **anticipated many features of the modern computer**, including a **central processing unit**, a **memory store** (with capacity for 1,000 fifty-digit numbers), and a **print-out facility**. It was intended to be **programmable by punched cards**. It was **never built** and was probably **too advanced for the cogwheel technology of the time**.

FIRST PROGRAMMER

Babbage's friend **Ada Lovelace** described algorithms that **could run on the proposed engine**, earning her the title of the **world's first computer programmer**.

SEED DRILL

*The signature technology of the Agricultural Revolution
was a simple but effective bit of engineering.*

MECHANICAL

HAND SOWING

Up until the widespread adoption of seed drills, almost all arable crops were **sown by hand**, which involved workers walking across a field and scattering seed. This was **wasteful and inefficient** for several reasons: **birds ate a lot of the seed** because it lay on top of the soil; plants might grow **too close together**, **lowering yields**; tasks such as **clearing weeds** and **applying fertilizer** took longer because plants grew haphazardly around the field.

DRILLING TALES

As early as 1500 BC, the ancient **Babylonians** were using a type of seed drill, while the **Chinese** invented **multi-tube iron drills** around the first century AD. Seed drills were developed further in **Renaissance Italy**.

JETHRO TULL

British agronomist **Jethro Tull** had observed farming practices on the continent and returned to his own farm **determined to make agriculture more efficient and yields higher**. In 1701 he adapted earlier ideas to create **his first seed drill**.

THE MECHANICAL SEED DRILL

Tull's device was a **horse-drawn carriage with a rotating cylinder containing seeds**. As the carriage rolled along, the box turned and **seeds fell out of slots into a hopper**, which **dropped them in furrows dug by the plough at the front of the carriage**. At the rear of the carriage was a **harrow** that **pushed soil on top of the seed**. By replicating the arrangement in parallel, a **single farmer could plant multiple parallel rows of seeds at the same time**.

THIS IS NOT A DRILL

No actual drilling was involved; Tull explained that he called the machine a seed drill because "**drilling**" was what farmers called **dropping seeds directly into a furrow**.

SEED SOLUTIONS

The seed drill **protected seeds from birds**, **improved** their **drainage** and **access to nutrients**, and **planted them in rows**, making the resulting crop rows **easier to weed** and otherwise **tend**, helping to increase the **efficiency** of agricultural labor.

Carries the seeds

Opens the furrow to a uniform depth

Places the seeds in the furrows

Covers the seeds by compacting soil around them

WEAVING MACHINES

*The British textile industry drove the start of the Industrial Revolution,
thanks to a sequence of engineering advances.*

COTTAGE INDUSTRY

In the early eighteenth century, **spinning** and **weaving**, the two stages of textile manufacture, were both **cottage industries**, performed by families in their own homes. **Productivity** was **limited by** factors such as **reliance on human power**, the **difficulty of spinning strong cotton thread**, and **narrow looms**.

WARP AND WEFT

Weaving cloth involves arranging threads in parallel lines (the **warp**) and then passing across the **weft** thread, over and under each warp thread. The **weft was threaded from a shuttle that was passed through by hand**. **Hand looms** could thus only be **as wide as the span of a man's reach**, and this **limited the width of cloth** that could be made at any one time.

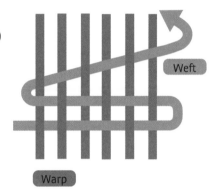

Weft

Warp

MECHANICAL

THE FLYING SHUTTLE

In 1733 **John Kay** invented a **shuttle that ran on little wheels** and could be **pulled back and forth by a cord**. This **flying shuttle** allowed much **wider looms** and **sped up the process** considerably, so that **one person could do the work of two or more**.

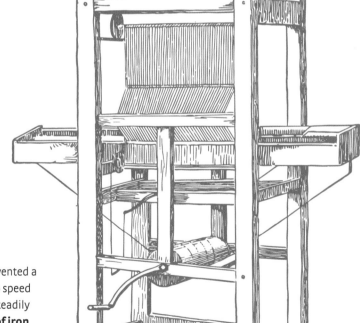

THE POWER LOOM

In 1785 clergyman **Edmund Cartwright** invented a **power loom**, which used external power to speed up the weaving process. His machine was steadily improved and **later versions were made of iron rather than wood**, so that they could be **powered by steam engines**.

SPINNING MACHINES

Engineering innovations triggered a kind of arms race between spinning and weaving, where each side strove to keep up with the supply/demand of the other, and thus drove further innovation.

THE SPINNING JENNY

Named after his daughter, the spinning jenny was invented in 1767 by English weaver **James Hargreaves**. **Spinning** was **done with a wheel** that **drove the rotation of a spindle**; Hargreaves reasoned that the **wheel could drive more than one spindle at a time**. The **spinning jenny had eight spindles**.

THE WATER FRAME

Richard Arkwright was an English entrepreneur who realized that even with **spinning jennies, spinners could not keep up with the demand for thread** stimulated by the **flying shuttle**. From 1764, he worked on a large **spinning machine**, called the **water frame** because it was driven by **water power**. The machines had to be housed in special buildings (factories), known as mills because, like grain mills, they used **water wheels**.

THE SPINNING MULE

Cotton thread could now be produced at a sufficient rate, but its **quality was variable**. In 1779 spinner **Samuel Crompton** combined the best features of the **spinning jenny** and the **water frame** to create the **spinning muie**, a device that **spun very fine and even thread**.

MECHANICAL

EVANS'S GRAIN MILL

Oliver Evans (1755–1819) was the preeminent US inventor of the late eighteenth century. His combination of innovations to produce an automated grain-milling production line transformed the milling industry.

AGAINST THE GRAIN

Engineering is all about **solving problems**. In 1782 Evans and his brothers took over running a grain mill, and he determined to engineer solutions to the problems he encountered:

Manual labor was used to **haul sacks of grain and flour up and down the mill**, and to spread out grain.

Dirt, **bugs**, and other **contaminants** got into the flour.

The **milling and drying process took a long time**, reducing efficiency.

The length of the process allowed more time for **grain and flour to spoil and pests to flourish**.

The **labor-intensive process** made milling **more costly**.

FIVE MACHINES

To solve these problems, Evans engineered a system that **combined previous innovations and introduced some new ones**. By 1787 he had designed a building that would run on **water power**, which would drive **five innovative devices**:

- A **bucket elevator**: a **series** of wooden or tin **scoops** on a **leather belt backing**, which **replaced the labor of men** hauling sacks up ladders.
- An **auger or screw conveyor**, to **shift material horizontally**.
- A "**drill**": an "**endless**" (i.e., looped) **conveyor belt** for **carrying material up an incline**.
- A **descender**: another endless conveyor belt to carry material **down**.
- A "**hopper-boy**": a machine taking its name from that given to the worker previously responsible for **spreading out the flour** on the floor to **cool and dry**, the hopper-boy had a **rotating rake** to spread flour.

EFFICIENCY GAINS

Evans's mill **enabled a single worker to do the work of five**, processing 300 bushels (approximately eight tons) of grain per hour.

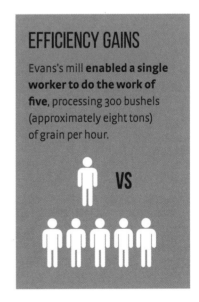

THE COTTON GIN

Eli Whitney's cotton-carding machine transformed the social economy of the American South and changed the course of history.

STICKY SEEDS

Eli Whitney was a Northerner who moved to Georgia in 1792, where he discovered that inland cotton plantation owners, able to grow only **short-staple cotton with sticky seeds**, were **struggling to turn it into a profitable product**. What was needed was a **quick**, **cheap way to remove the seeds**.

BRUSH OFF

Whitney's design was simple and elegant in concept. A **roller covered in little hooks caught cotton fibers fed into a hopper** and **dragged them through a fine-toothed comb-like grid**. The **cotton fibers could pass through**, but the **seeds could not**. On the other side of the grid, a set of **revolving bristles brushed the cotton fibers off the roller** so that it could **revolve to collect more**.

HORSE POWER

Whitney devised the machine so that it could be **powered by a horse pulling a wheel around**. "One man and a horse will do more than fifty men with the old machines," he wrote to his father.

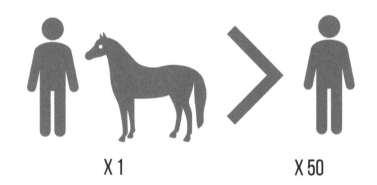

X 1   X 50

BAD NEWS

Thanks to the gin and other innovations, **cotton growing became immensely profitable**, spurring a colossal **increase in the scale of cotton agriculture** and the **demand for slaves**, who were driven ever more mercilessly in an **industrialized system of exploitation**. The **South was now irrevocably tied to slavery**.

PORTSMOUTH BLOCK MILLS

Celebrated as one of the earliest industrial mass production plants in the world, the Portsmouth block mills were the brainchild of Marc Brunel.

BLOCK PARTY

At the end of the eighteenth century, amid the tumult of the **Napoleonic Wars**, the British Royal Navy (RN) had become a colossal and technologically advanced enterprise, gobbling up material and equipment, including blocks, the **wooden parts of block-and-tackle assemblages used in pulleys**. A typical **large ship needed 1,000 blocks** of various sizes, and the **RN needed over 100,000 blocks a year**.

INDUSTRIAL HEARTLAND

Under the leadership of **Brigadier-General Sir Samuel Bentham**, Inspector General of Naval Works, the shipyards at Portsmouth were transformed into one of the world's leading industrial sites, with **cutting-edge technology** and **advanced workshops**. On top of a **huge drainage basin**, Bentham had **constructed a complex of wood mills**, with **machinery driven by steam engines**.

RISE OF THE MACHINES

Working with engineers such as **Henry Maudslay**, **Simon Goodrich**, and **Bentham** himself, **Brunel** perfected a set of machines—at **first made of wood** and **later of metal**—that would **cut wood into standardized blocks** and then **shape these** through **drilling, boring, cutting, milling**, and **planing**, to produce blocks of **various size** and **specification**.

BRUNEL'S PROPOSAL

Bentham himself had designed some machines to **speed the production of blocks**, which up to this point had been cut by hand, but his plans had not advanced. In 1802 **Marc Brunel** presented to the RN his **designs for a set of machines to make blocks**, and Bentham pushed for their adoption.

PRODUCTION BOOST

The machines **allowed ten men to produce as many blocks as 110 skilled woodworkers**. By 1808, the block mills were turning out **130,000 blocks a year**.

STIRLING ENGINES

Closed, quiet, low-temperature engines that run on heat differentials, Stirling engines were invented in 1816 but are only now coming into their own.

BOILING POINT

Robert Stirling was a Scottish clergyman in an industrial region, concerned about the **danger to his parishioners from steam engines**. These used **boilers** to make steam, and the **high pressures** involved frequently over-taxed the weak iron boilerplate, **causing explosions**.

AIR ENGINES

Stirling had heard of **air engines**, which **used air instead of steam as the medium that pushed the pistons and turned heat into work**, but he knew they were **ineffective** and he **determined to improve them**. By 1816, he had invented his **Heat Economizer**, a **heat exchange device**, a.k.a. a **regenerator**, which greatly **improved the efficiency of air engines**.

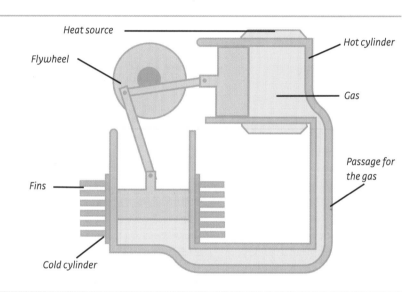

Heat source

Flywheel

Hot cylinder

Gas

Fins

Passage for the gas

Cold cylinder

MECHANICAL

SAFER ENGINES

The following year, Stirling and his brother incorporated his economizer into an **air engine**, producing an engine that **operated at lower heat than a steam engine**, **could not cause steam scalding** (because it didn't use steam), and **could not explode** (because there was no high-pressure boiler). By 1818, a Stirling engine was **installed to operate a pump at a nearby quarry**.

MODERN USES

Stirling engines were **expensive** and **failed to displace steam engines**, especially when advances in **steel making** produced **stronger boilerplates** and steam engines **became safer**. But the **principles**, **efficiency**, and **very low noise levels** of Stirling engines have seen them find **specialist applications in submarines** (which try to stay quiet) and **eco-friendly power generation**.

STEAM HAMMERS

An icon of the industrial era, the steam hammer harnessed the power of steam in the service of both brute force and finesse.

TILT HAMMERS

Steam power had been pressed into the **blacksmith's** service before the creation of the steam hammer. The **tilt or trip hammer** was essentially a **mechanized blacksmith's arm**, with a **hammer on the end of an arm**. The arm was **raised by water or steam power** and then **released to fall in an arc**.

WIDE LOAD

In 1839 tool-making engineer **James Nasmyth** received a letter from a top engineer at the **Great Western Steam Company**, lamenting the fact that he was **unable to find anyone who could forge a massive axle for the planned giant paddle wheels of the** SS *Great Eastern* (the design was later changed to **screw propellers**). **Tilt hammers** would **not** be **able to do the job** because the sheer size of the piece blocked their arc of descent.

NASMYTH'S HAMMER

Nasmyth immediately sketched out a **new kind of hammer** that would do the job, describing the elements thus: "a massive anvil . . . a block of iron constituting the hammer . . . and an inverted steam cylinder to whose piston-rod the hammer-block was attached." **Steam power raised the piston and the attached hammer, while letting out the steam let the hammer fall under gravity.**

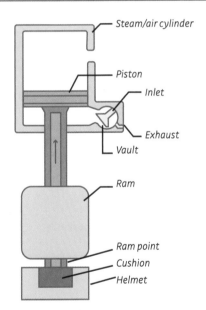

Steam/air cylinder
Piston
Inlet
Exhaust
Vault
Ram
Ram point
Cushion
Helmet

BIG HAMMER

Nasmyth **patented his Steam Hammer design** in 1842, by which time French industrialists at the manufactories in **Le Creusot** had already built a working version. Le Creusot would later host a **colossal 110-ton version**, which **stands today as a monument**.

GENTLE TOUCH

The **cylinder of Nasmyth's hammer** had a **valve** that **regulated the release of steam**, allowing for **very delicate control**. His favorite demonstration was to **break an eggshell in a wineglass without damaging the glass**, before **pounding molten ingots into sheets**.

ROBOTS

"Robot" can have many meanings, but it is generally taken to be an autonomous machine.

AUTOMATONS

Mechanics in **ancient Europe** and the **medieval Islamic world** made **ingenious automata** to **amuse**, **impress**, and **provoke debate**. The practice continued into the modern world. The **Digesting Duck** created by **French engineer Jacques de Vaucanson in 1738** was said to be able to **flap its wings**, **quack**, **eat**, and even **defecate**.

WORKER BOTS

Ancient Greek myth told of mechanical servants and workers, a theme taken up by science fiction at least as far back as the nineteenth century, so it was expected that real-life technology would soon catch up. In fact, the **first industrial robot did not go into service until 1961**.

UNIMATE

Unimation was a company formed in 1956 by engineer **Joseph Engelberger** and inventor **George Devol**, who created a **programmable automation system**. Together they developed the Unimate, a **programmable robot arm that went into service in a General Motors factory** in 1962.

MOWERS AND VACUUM CLEANERS

Outside of the assembly line, **few robots have become commercially available**. Puck-like robots that **mow grass** or **clean floors** are commonplace, as are **robot toys**. Self-driving cars are **effectively robots**.

ROBOT CHALLENGES

It has proven hard to get robots out of the laboratory because of a number of **engineering challenges yet to be overcome**. These include **sufficient power for autonomous operation**; **adequate AI**, which is necesary to effect real-time operation of tasks that are easy for humans, such as **visual recognition** or **navigating messy environments**; and **robust and inexpensive construction**—robots are **too fragile and valuable for real-world operation**.

JAMES DYSON

Famous for his revolutionary vacuum cleaner design, British engineer James Dyson is a fearless inventor and iconoclast.

BALL BARROW

Dyson went to **art school** and then **studied design**, before becoming an engineer and inventor. One of his earliest innovations was the **ball barrow**: a wheelbarrow where the **wheel** has been **replaced by a large ball**, **increasing maneuverability**.

DUST BUSTER

At his factory for making the ball barrow, Dyson encountered a **problem with dust** from the resin used to coat parts of the barrow. He installed an **industrial cyclone tower**, which uses **spinning air to remove dust particles**.

THIS VACUUM SUCKS

In 1978 Dyson became **frustrated with the family vacuum cleaner**, finding that it was **losing suction power** because the **bag became clogged with dust**. He had the idea to try **applying cyclonic technology in a domestic vacuum cleaner**.

DEVELOPMENT HELL

Over the next five years, Dyson developed **5,127 prototypes**, which he **built himself** out of **brass**, **aluminum**, and **Perspex**. Attempts to partner with established manufacturers failed, so Dyson **started his own company**, which now has **sales of around $6 billion a year and employs 5,800 engineers around the world**.

CREATIVE THINKING

Dyson and his team of engineers went on to **design creative engineering solutions for other fields**, producing new designs for **hand dryers**, **fans**, and **hair dryers**. He is also working on an **electric car**.

TIMELINE

Dates are taken from when a significant advance was made, rather than indicating the earliest known version of a technology or engineering feat.

c.60,000 BC	Beam
c.60,000 BC	Column
c.60,000 BC	Truss construction
c.11,000 BC	Bioengineering
c.10,000 BC	Ships
c.9000 BC	Rockets
c.7000 BC	Bow and arrow
c.4500 BC	Bronze Age
c.4000 BC	Roads
c.3800 BC	The wheel
c.3000 BC	Surveying
c.2900 BC	Dams
c.2500 BC	Toilets
c.2500 BC	Arches
1750 BC	Risk
c.1300 BC	Bridges
c.1200 BC	Iron
c.850 BC	Siege engines
660 BC	Lighthouses
c.550 BC	Tunnels
c.550 BC	Water supplies and sewers
c.550 BC	Crossbow
515 BC	Canals and locks
c.500 BC	Vulnerability (origin of engineering earthquake-resistant structures)
c.400 BC	Renewable energy
c.400 BC	Waterwheels
c.350 BC	Trebuchet
c.300 BC	Energy
c.220 BC	Archimedes (his major work)
c.200 BC	Ancient machines
200 BC	Levers
c.100 BC	Ballista
80 BC	Antikythera mechanism

C.AD 1	The Aeolipile
c.120	Domes
135	Zhang Heng
644	Windmills
850	Gunpowder
1280	Spectacles and lenses
1326	Early cannons
1364	Firearms and artillery
1500	Great Wall of China
c.1500	Leonardo da Vinci (his major engineering work)
c.1600	Mechanics
1642	Calculating machines
1645	Pumps
1657	Clocks
1660	Electrostatic generators
1675	Elasticity
1701	Seed drill
1712	Steam engines
1733	Weaving machines
1745	Leyden jar
1767	Spinning machines
1771	Factories
1773	Soil mechanics
1776	Cannons
1783	Steamships
1787	Evans's grain mill
1793	Cotton gin
1800	Voltaic pile
c.1800	Geoengineering
1801	Suspension bridges
1801	Locomotives
1802	Portsmouth block mills
1809	Early electric lights
1817	Stirling engines
1823	Babbage's engines
1824	Thermodynamics
1824	Cement

1825	Tunneling shield	1906	Triode valve
1825	Resistors	1913	Ford and the assembly line
1826	Omnibuses	1916	Tanks
1829	Stephenson's Rocket	1926	Television
1832	Concrete	1928	Iron lung
1832	Dynamos	1930	Jet engine
1833	Isambard Kingdom Brunel	1939	Helicopters
1837	Breech-loading artillery	1940	Radar
1837	Telegraph	1940s	Finite elements analysis
1839	Steam hammers	1943	Bouncing bomb
c.1840	Viscosity	1945	Atom bomb
1840	Heat	1946	Microwave oven
1852	Airships	1947	Transistor
1856	Bessemer's converter	1948	Information theory
1858	Sir Joseph Bazalgette (construction of London's sewers)	1952	Artificial heart valves
		1952	Pacemaker
1861	Bicycles	1953	Heart-lung machine
1861	Elevators	1957	Cochlear and retinal implants
1866	Electric railways	1957	Sputnik
1866	Self-exciting dynamo	1958	Integrated circuit
1868	Control theory	1959	Hovercraft
1868	Powered flight	1960	Laser
1870	Skyscrapers	1962	Robots
1875	Emergence	1968	Human–computer interaction
1876	Internal combustion engine	1969	Apollo program
1876	Telephone	1972	Genetic engineering
1877	Phonograph	1974	Personal computers
1879	Lightbulb	1978	Global positioning system
1881	Nikola Tesla (invention of the polyphase motor)	1981	Nanotechnology
		1983	James Dyson
1881	Polyphase induction motor	1988	Tissue engineering
1882	Electric power generation	1990	Saving the Leaning Tower
1884	Steam turbines	1990	Hubble Space Telescope
1885	Automobiles	1990	Search engines
1885	Machine guns	1993	Bionics
1886	Bone repair	1994	Channel Tunnel
1886	AC vs DC	1998	International Space Station
1889	Eiffel Tower	1998	Quantum computing
1890	Artificial joints	2002	Elon Musk (founding of SpaceX)
1895	Medical imaging	2010	Artificial life
1898	Submarines	2013	CRISPR-Cas9
1898	UAV drones	2014	Self-driving cars
1901	Radio	Future	Space elevator
1903	ECG	Future	Dyson sphere
1903	Wright brothers	Future	Future weapons
1904	Diode valve	Future	Artificial intelligence

GLOSSARY

aggregate: coarse- to medium-grained particulate material—bigger grains than powder but smaller than rocks

alloy: a mixture of metals

arch: a curved structure used to span an opening and to support loads from above

ballistics: science of the motion of projectiles

bioengineering: a.k.a. biomedical engineering—application of engineering principles and techniques to biology and medicine

biomimicry: copying or drawing inspiration from natural engineering

capacitor: a device for storing electric charge in an electrostatic field

carburization: adding carbon to iron to produce steel

cement: a binding substance used in construction

compression: a pushing or squeezing force

concrete: artificial rock or stone

current: the rate of flow of electric charge

dome: a form of vault; an arch that is deeper than its span

drag: a type of friction; force acting opposite to relative motion of an object moving with respect to a surrounding fluid

ductile: bendy—opposite of brittle

dynamo: an electricity generator that outputs direct current

electric charge: the quantity of unbalanced electricity in a body (either positive or negative)

electromagnetic induction: the phenomenon whereby moving magnetic fields can induce electrical fields and vice versa

electrostatic charge: electric charge that is not moving

emergence: the principle that systems can exhibit properties that result from the interaction of elements but that are not present or could not be predicted from those elements in isolation

energy: the capacity for doing work

engine: a machine with moving parts that converts power into motion

equilibrium: differing variables (such as forces) that balance out to produce no overall change

factory: a system for manufacturing; normally understood to be a building or set of buildings

field: a region in which each point is affected by a force

force: a push or pull

friction: the force resisting relative motion of surfaces or layers sliding against one another

gear: a device for transmitting rotary motion

geoengineering: large-scale interventions to change climate or other global-scale natural processes

hydraulic: of or concerning water

isotopes: forms of an element that differ in terms of how many neutrons are in the nucleus, which can affect the stability of the nucleus

lever: a type of simple machine: a device for doing work by amplifying force

lift: lifting force

load: a collection of forces acting on an object, especially weight, or a source of pressure

mechanics: the science of bodies in motion or equilibrium; also, the study in Classical times of the principles of technology

metallurgy: the art and science of working with metals

motor: a machine that supplies motive power for a vehicle or other device with moving parts

nanotechnology: materials, structures, and devices constructed and/or operating at the scale of molecules and atoms

parabolic: having the form of a parabola, a kind of curve described by projectiles falling under gravity

pneumatic: of or concerning gases, including air

power: energy or work per unit time; rate at which work is done or energy converted

projectile: a moving body acted on only by gravity

prosthetic: an artificial body part or a body-part substitute

rectification: the process of converting AC into DC or modifying electromagnetic wave forms

resistance: the tendency of a material to resist the passage of an electric current and to convert electrical energy into heat energy

semiconductor: material that can switch between being a conductor and being an insulator, depending on external factors

shear: bending force; unaligned forces pushing in opposite directions

stress: force intensity or force per unit area; internal forces acting in a body

systems engineering: the study and practice of putting elements together so that they interact as a system

tensile strength: the amount of tension a material can sustain before failure (breaking)

tension: pulling force

thrust: the force driving a body forward

torque: rotational force

torsion: twisting around the long axis of a body while one end of the body is fixed

transistor: a contraction of "trans-resistor": a kind of electronic valve

turbine: a machine with a rotor, usually with vanes or blades, that extracts energy from fluid flow

work: moving something against a force

FURTHER READING

Roma Agrawal, *Built: The Hidden Stories Behind Our Structures*, Bloomsbury, 2018

J. E. Gordon, *Structures: Or Why Things Don't Fall Down*, DaCapo Press, 2003

Adam Hart-Davis, *Engineers: From the Great Pyramids to Spacecraft*, Dorling Kindersley, 2017

J. L. Heilbron (ed.), *Oxford Companion to the History of Modern Science*, OUP, 2003

J. G. Landels, *Engineering in the Ancient World*, Constable, 2000

Joel Levy, *50 Weapons that Changed the Course of History*, Firefly, 2012

David Macaulay, *The Way Things Work*, Dorling Kindersley, 1990

Donald A. Norman, *The Design of Everyday Things*, MIT Press, 2013

Henry Petroski, *The Evolution of Useful Things*, Vintage, 1997

Adam Piore, *The Body Builders: Inside the Science of the Engineered Human*, Ecco Press, 2017

Sal Restivo (ed.), *Science, Technology, and Society: An Encyclopedia*, OUP, 2005

L. T. C. Rolt, *Victorian Engineering: A Fascinating Story of Invention and Achievement*, Penguin, 2000

Robert Temple, *The Genius of China: 3,000 Years of Science, Discovery, and Invention*, Inner Traditions, 2007

Simon Winchester, *Exactly: How Precision Engineers Created the Modern World*, William Collins, 2019

WEBSITES

American Society of Mechanical Engineers: asme.org

Edison Tech Center: edisontechcenter.org

Grace's Guide to British Industrial History: gracesguide.co.uk

Institute of Civil Engineers, ICE 200: www.ice.org.uk/what-is-civil-engineering/what-do-civil-engineers-do

Transistorized! The Story of Microelectronics: www.pbs.org/transistor/index.html